〔大数据技术丛书〕

Hive

入门与大数据分析实战

迟殿委 著

清華大學出版社
北京

内 容 简 介

Hive 是基于 Hadoop 的一个数据仓库工具，用来进行数据的提取、转换、加载，这是一种可以存储、查询和分析存储在 Hadoop 中的大规模数据的机制。Hive 能将结构化的数据文件映射为一张数据库表，并能提供 SQL 查询分析功能，将 SQL 语句转换成 MapReduce 任务来执行，从而实现对数据进行分析的目的。本书配套示例源码、PPT 课件、教学大纲。

本书共分 11 章，内容包括数据仓库与 Hive、Hive 部署与基本操作、Hive 语法基础、Hive 数据定义、Hive 数据操作、Hive 查询、Hive 函数、Hive 数据压缩、Hive 调优、基于 Hive 的网站流量分析项目实战、旅游酒店评价大数据分析项目实战。最后的两个项目实战（均包括 SQL 和 Java 编程两种解决方法）帮助读者提高 Hive 大数据分析的综合实战能力。

本书可作为Hive数据仓库初学者的入门书，也可作为Hive大数据分析与大数据应用开发工程师的指导手册，还可作为高等院校或者高职高专计算机技术、人工智能、大数据技术及相关专业的教材或教学参考书。

图书在版编目（CIP）数据

Hive 入门与大数据分析实战 / 迟殿委著. 一北京：清华大学出版社，2023.5（2024.2重印）
（大数据技术丛书）
ISBN 978-7-302-63421-8

Ⅰ. ①H… Ⅱ. ①迟… Ⅲ. ①数据库系统一程序设计 Ⅳ. ①TP311.13

中国国家版本馆 CIP 数据核字（2023）第 079534 号

责任编辑：夏毓彦
封面设计：王　翔
责任校对：闫秀华
责任印制：杨　艳

出版发行：清华大学出版社
　　网　　址：https://www.tup.com.cn，https://www.wqxuetang.com
　　地　　址：北京清华大学学研大厦 A 座　　　邮　　编：100084
　　社 总 机：010-83470000　　　邮　　购：010-62786544
　　投稿与读者服务：010-62776969，c-service@tup.tsinghua.edu.cn
　　质 量 反 馈：010-62772015，zhiliang@tup.tsinghua.edu.cn

印 装 者：三河市铭诚印务有限公司
经　　销：全国新华书店
开　　本：190mm×260mm　　印　　张：14　　字　　数：377 千字
版　　次：2023 年 6 月第 1 版　　印　　次：2024 年 2 月第 2 次印刷
定　　价：69.00 元

产品编号：102320-02

前　言

如今各个行业都积累了海量的历史数据，并不断产生大量的新数据，数据计量已经发展到PB、EB、ZB、YB，甚至BB、NB、DB级别。由此催生了一门全新的技术——Hive离线计算。Hive是Hadoop生态体系的关键组件之一，它的出现使得海量数据可以继续使用传统的数据分析方法SQL语句来处理，降低了数据分析人员的学习成本。数据分析人员不需要学习新的脚本语言，可以继续使用熟悉的SQL结构化查询语句来分析大规模数据。但是，Hive的SQL语句不再运行在传统的数据库或者数据仓库中，而是运行在大数据分布式并行计算处理平台上。

本书内容

本书内容按照从易到难、理论与实战相结合的思路来组织。俗话说“工欲善其事，必先利其器”，本书在介绍数据仓库和Hive的基本概念之后，马上开始讲解从创建虚拟机、安装Linux操作系统到逐步完成Hive部署的详细过程；然后在部署完成的Hive环境基础上，学习Hive语法基础、Hive数据定义语言、Hive数据操纵语言、Hive数据基本查询等相关操作；接下来深入介绍Hive的其他功能，包括Hive函数、Hive数据压缩、Hive调优等；最后，本书通过网站流量分析项目实战、旅游酒店评价大数据分析项目实战这两个开发案例，帮助读者提升大数据分析的综合实战能力。这两个实战项目都给出了SQL实现和Java编程实现这两种解决方法，为读者做大数据开发起到抛砖引玉的作用。

本书目的

本书目的是带领读者系统掌握Hive大数据分析工具的使用与开发方法，并通过两个综合项目案例帮助读者提高Hive大数据分析的实战能力。

配套示例源码、PPT课件

本书配套示例源码、PPT课件、教学大纲，需要用微信扫描右边二维码获取。如果阅读中发现问题或疑问，请联系booksaga@163.com，邮件主题写“Hive入门与大数据分析实战”。

本书适合的读者

本书可作为Hive数据仓库初学者的入门书、Hive离线大数据分析人员的参考手册，也可作为高校开设大数据平台搭建、数据仓库技术或大数据开发课程的参考教材。

学习本书要求读者有一定的Java编程基础并了解Linux系统的基础知识。本书每一个章节的实践操作都有详细清晰的步骤讲解，即使读者没有任何大数据基础，也可以对照书中的步骤成功搭建属于自己的大数据分析平台；可以说本书是一本真正能提高读者动手能力、以实操为主的Hive入门书。通过本书的学习，结合每章的示例源代码，读者能够迅速理解和掌握Hive技术框架，并能熟练使用Hive数据仓库进行大数据分析和大数据应用开发。

笔　者

2023年3月

目　　录

第1章

数据仓库与 Hive

1.1　数据仓库概述

数据仓库是在数据库的基础上建立起来的，但与传统的数据库又有较大的不同，它将分布在不同数据库中的数据集成起来，将转换后的关系型数据及其他复杂类型数据存储成为一种面向分析的数据集合。

1.1.1　数据仓库特征与重要概念

1. 数据仓库一般具有的特征

1）数据仓库的数据是面向主题的

主题是一个抽象的概念，是在较高层次上综合、归纳企业信息系统中的数据并进行分析利用的抽象。在逻辑意义上，它对应着企业中某一宏观分析领域所涉及的分析对象。

2）数据仓库的数据是集成的

在数据进入数据仓库之前，必然要经过加工与集成，对不同的数据来源统一数据结构和编码，将原始数据由面向应用转向面向主题。

3）数据仓库的数据是可更新的

数据仓库的数据主要供企业决策分析之用，所涉及的数据操作主要是数据查询，一般情况下并不进行修改操作，因而数据经集成进入数据仓库后是极少或根本不更新的。

4）数据仓库的数据是随时间变化的

数据仓库中的数据不可更新是针对应用来说的，也就是说，数据仓库的用户在进行分析处理时是不进行数据更新操作的，但并不是说，在从数据集成输入数据仓库开始到最终被删除的整个数据生存周期中，所有的数据仓库都是永远不变的。数据仓库内的数据时限一般在 5～10 年，故数据的编码包含时间项。数据仓库要周期性地收集和整理数据，以适应决策支持系统。

2. 数据仓库中的几个重要概念

下面再介绍一下数据仓库中的几个重要概念：粒度、分割、维、元数据。

1）粒度

粒度是指数据仓库中数据单元的详细程度和级别。一般操作型系统中处理的数据都是详细数据，其粒度是最低的。但在分析型处理中需要通过数据的概括和聚集形成较高粒度的数据。粒度越小，细节程度越高，级别就越低；反之，数据的综合程度越高，粒度越大，级别就越高。数据的粒度越高，所需要存储的数据量越少，但对决策者的重要性却随之增加，且能够为用户提供快速方便的查询。数据仓库一般提供多种粒度的数据，不同粒度的数据用于不同类型的处理。比如销售产品，其粒度可以是每天的数据，也可以是每周、每月、每季度，甚至每年记录统计的数据。通常的数据粒度有详细数据、轻度综合、高度综合三级。

2）分割

分割是指将逻辑上统一的数据分割成较小的、可以独立管理的物理单元进行存储，以便于提高数据处理效率，数据分割后的单元称为分片。数据分割的标准是按照实际情况确定的，通常按日期、地理分布、业务范围等进行分割。数据分割后较小单元的数据处理相对独立，使得数据更易于重构、索引、恢复和监控，处理起来更快。比如产品销售数据可以按照不同地域（如北京地区、东北地区、华北地区等）进行分割。

3）维

维是人们观察数据的特定角度，是数据的视图。比如可以从销售时间、销售地区分布等不同角度来观察产品销售数据。维可以有细节程度的不同描述方面，这些不同描述方面称为维的不同维层次。最常用的维是时间维，时间维的维层次可以有日、周、月、季、年等。数据仓库中的数据按照不同的维组织起来形成一个多维立方体。

4）元数据

所谓元数据就是关于数据的数据，它描述了数据的结构、内容、码、索引等项内容。传统数据库中的数据字典就是一种元数据，但在数据仓库中，元数据的内容比数据库中的数据字典更丰富、更复杂。

1.1.2 数据仓库的数据存储方式

数据仓库的数据存储方式一般说来有两种，即基于关系表的存储方式和基于多维数据库的存储方式。

1. 基于关系表的存储方式

基于关系表的存储方式是将数据仓库的数据存储在关系数据库的表结构中，在元数据的管理下完成数据仓库的功能。这种组织方式在建库时有两个主要过程用以完成数据的抽取：首先要提供一种图形化的点击操作界面，使分析员能对源数据的内容进行选择、定义多维数据模型；然后再编制程序，把数据库中的数据抽取到数据仓库的数据库中。基于关系表的数据存储方式主要有星型模型和雪花模型两种。

1）星型模型

大多数数据仓库都采用“星型模型”来表示多维概念模型。数据库中包括一张“事实表”，对于每一维都有一张“维表”。“事实表”中的每条元组都包含指向各个“维表”的外键和一些相应的测量数据。“维表”中记录的是有关这一维的属性。销售数据仓库的星型模型图如图 1-1 所示。

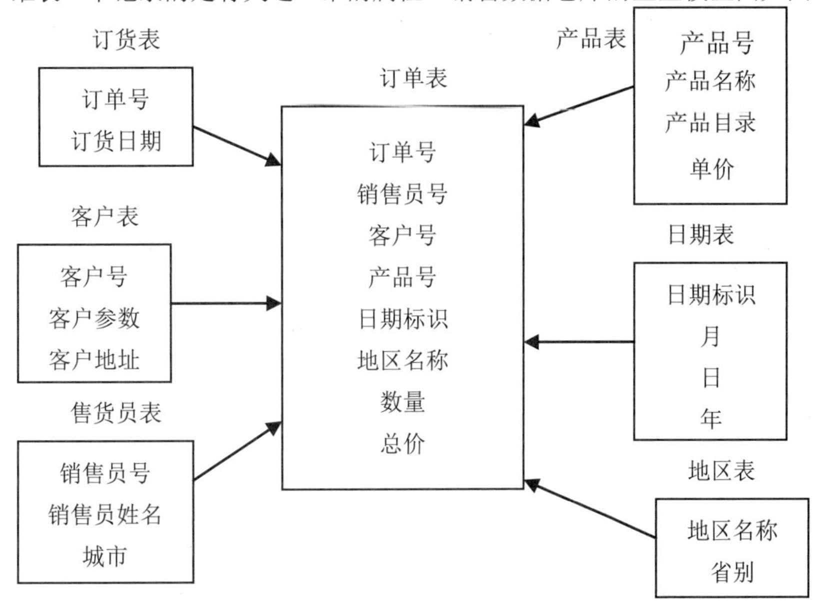

图 1-1　销售数据仓库的星型模型

事实表中的每一个元组包含一些指针（是外键，主键在其他表中），每个指针指向一张维表，这就构成了数据库的多维联系。相应每个元组中多维外键限定数据测量值。在每张维表中，除包含每一维的主键外，还有说明该维的一些其他属性。维表记录了维的层次关系。在数据仓库模型中执行查询的分析过程，需要花费大量时间在相关各表中寻找数据。而星型模型使数据仓库的复杂查询可以直接通过各维的层次比较、上钻、下钻等操作完成。

星型模型的数据组织方式存在数据冗余、多维操作速度慢的缺点，但这种方式是主流方案，大多数数据仓库集成方案都采用这种方式。

2）雪花模型

“雪花模型”是对星型模型的扩展，它进一步层次化了星型模型的维表，原有各维表可能被扩展为小的事实表，形成一些局部的“层次”区域。对应的雪花模型图如图 1-2 所示。

雪花模型的优点：通过最大限度地减少数据存储量及联合较小的维表来改善查询性能。

雪花模型增加了用户必须处理的表数量，增加了某些查询的复杂性，降低了系统的通用程度，但同时这种方式可以使系统进一步专业化和实用化。前端工具仍然要用户在雪花的逻辑概念模式上操作，然后将用户的操作转化为具体的物理模式，从而完成对数据的查询。

从功能结构的划分来看，数据仓库系统至少应包含数据获取（Data Acquisition）、数据存储（Data Storage）、数据访问（Data Access）三个核心部分。

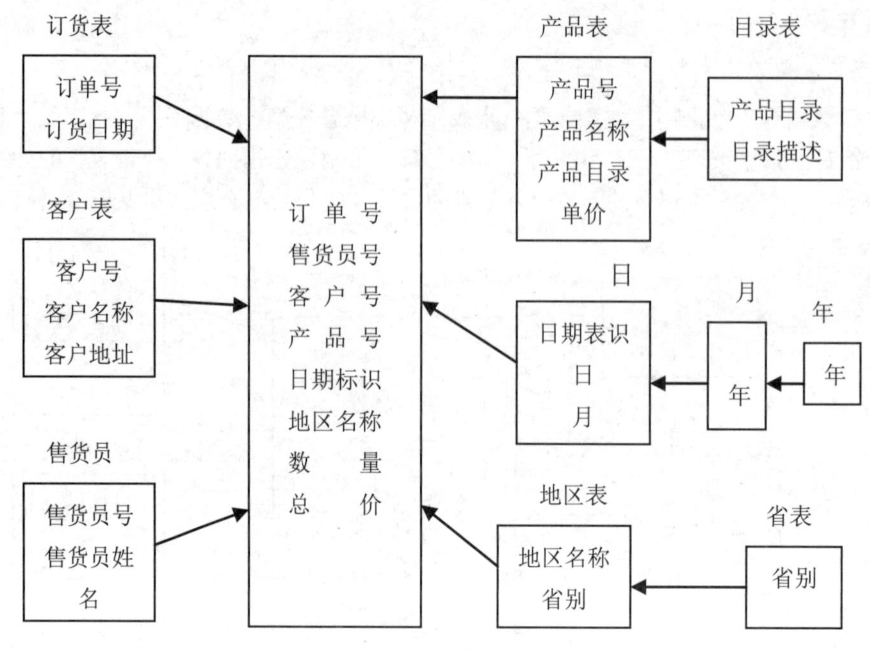

图 1-2　雪花模型

数据源是数据仓库系统的基础，是整个数据仓库系统的数据源泉。数据通常存储在关系数据库中，比如 Oracle 或者 MySQL。数据也可能来自文档资料，比如 CSV 文件或者 TXT 文件。数据还可能来自一些其他的文件系统。数据库是整个数据仓库系统的核心，是数据存放的地方，并能提供对数据检索的支持。

对不同的数据进行抽取（Extract）、转换（Transform）和装载（Load）的过程，也就是通常所说的 ETL 过程。

抽取是指把数据源的数据按照一定的方式从各种各样的存储方式中读取出来。对各种不同数据存储方式的访问能力是数据抽取工具的关键。因为不同数据源的数据格式可能会有所不同，不一定能满足业务的需求，所以还要按照一定的规则进行转换。数据转换包括：删除对决策应用没有意义的数据，将数据转换为统一的数据名称、定义及格式，计算统计和衍生数据，为缺失值赋予默认值，把不同的数据定义方式进行统一。只有转换后符合要求的数据才能进行装载。装载就是将满足格式要求的数据存储到数据仓库中。

2. 基于多维数据库的存储方式

多维数据的组织包括维数据组织和度量数据组织两个方面。维数据组织主要是组织多维数组数据结构和存储维的结构信息，度量数据则是以提高聚集查询的效率来进行组织的。

1）维数据组织

组织维数据首先要对维成员层次进行组织，然后将维成员映射为多维数组的坐标值。

2）度量数据组织

OLAP 操作通常需要处理整个多维数组，由于数据仓库的数据是海量数据，多维数组中的记录数一般很大，经常会超出系统内存，因此需要将多维数组进行分块（Chunk）。OLAP 操作以 I/O 块大小为单位进行存取，这样可显著提高数据访问的性能。

1.2　Hive 数据仓库简介

Hive 是基于 Hadoop 的一个数据仓库工具，用来进行数据的提取、转化、加载，这是一种可以查询和分析存储在 Hadoop 中的大规模数据的机制。Hive 数据仓库工具能将结构化的数据文件映射为一张数据库表，并提供 SQL 查询功能，能将 SQL 语句转换成 MapReduce（简称 MR）任务来执行。

关于 Hive 的描述可以归结为以下几点：

- Hive 是工具。
- Hive 可以用来构建数据仓库。
- Hive 具有类似 SQL 的操作语句 HQL。
- Hive 是用来开发 SQL 类型脚本，用于开发 MapReduce 操作的平台。

Hive 最初由 Facebook 开源，用于解决海量结构化日志的数据统计分析，是建立在 Hadoop 集群的 HDFS 上的数据仓库基础框架，其本质是将类 SQL 语句转换为 MapReduce 任务来运行。可以通过类 SQL 语句快速实现简单的 MapReduce 统计计算，十分适合数据仓库的统计分析。

所有 Hive 处理的数据都存储在 HDFS 中，Hive 在加载数据过程中不会对数据进行任何修改，只是将数据移动或复制到 HDFS 中 Hive 设定的目录下，因此 Hive 不支持对数据的改写和添加，所有数据都是在加载时确定的。

Hive 总体来说具有以下特点：

（1）Hive 是一个构建在 Hadoop 上的数据仓库框架。

（2）Hive 设计的目的是让精通 SQL 技能、但 Java 编程技能相对较弱的数据分析师能够快速进行大数据分析项目的开发与应用。

非结构化数据分析步骤如图 1-3 所示，其中 Hive 的能力在于直接分析通过 ETL 清洗过后的半结构化数据。

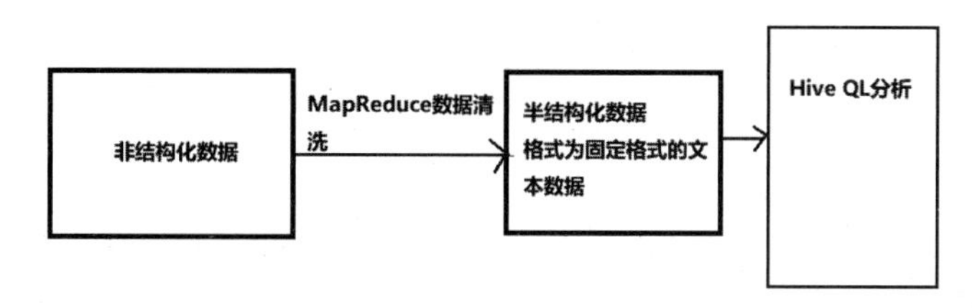

图 1-3　非结构化数据分析步骤

1.3 Hive 版本和 MapReduce 版本的 WordCount 比较

1. MapReduce 版本的 WordCount

读者之前应该都学习过 Hadoop 的 MapReduce 框架应用的相关知识，因此这里就不写完整的 MapReduce 应用的代码了。纯使用 MapReduce 方式的整个流程比较复杂，如果需要修改某个部分，那么首先需要修改代码中的逻辑，然后把代码打包上传到某个可访问路径上（一般就是 HDFS），再在调度平台内执行。如果是改动较大的情况，则可能还会需要在测试环境中多次调试。总之，就是会花比较多的时间在非业务逻辑改动的工作上。本节用于说明 Hive 和 MapReduce 二者的主要区别，具体的操作验证可在第 6 章之后进行。

2. Hive 版本的 WordCount

使用 Hive 来开发一个 WordCount 程序的基本流程如下：

步骤01 创建表：

```
hive> create table docs(line string);
OK
Time taken: 0.232 seconds
```

步骤02 导入数据（首先在/home/hadoop 路径下创建一个文本文件 derby.log，hadoop 是笔者的 CentOS 系统用户名，这个用户名在安装 Linux 时创建的）：

```
hive> load data local inpath '/home/hadoop/derby.log' into table docs;
Loading data to table default.docs
Table default.docs stats: [numFiles=1]
OK
Time taken: 1.119 seconds
```

步骤03 编写 Hive 版本的 WordCount，开发极其简单：

```
hive> create table word_count as
    > select word,count(1) as count from
    > (select explode(split(line,'\s+')) as word from docs) w group by word
    > order by word;
```

上面语句直接将查询运算的结果保存到 word_count 表中去。

由此可以看出 Hive 和 MapReduce 二者的主要区别在于下面两点。

1）运算资源消耗

从时间、数据量、计算量上来看，一般情况下 MapReduce 都是优于或者等于 Hive 的。MapReduce 的灵活性毋庸置疑。在转换到 Hive 的过程中，会有一些为了实现某些场景的需求而不得不用多步 Hive 来实现的情形。

2）开发成本/维护成本

毫无疑问，Hive 的开发成本远低于 MapReduce。后面会介绍，如果能熟练地运用 udf 和 transform，那么 Hive 的开发效率会更高。另外，由于使用了 SQL 语法对数据进行操作，因此处理起来非常直观，也让 Hive 开发更加容易上手。

1.4　Hive 和 Hadoop 的关系

Hive 构建在 Hadoop 之上，二者的关系示意图如图 1-4 所示。

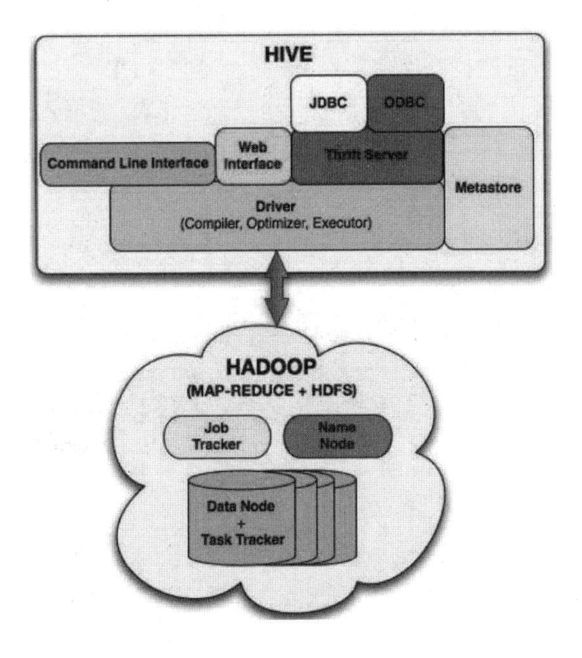

图 1-4　Hive 与 Hadoop 关系

它们关系解释如下：

- Hive 对外提供 CLI、Web Interface（Web 接口）、JDBC、ODBC 等访问接口，Hadoop 提供后台存储和计算服务。
- HQL 中对查询语句的解释、优化、生成查询计划都是由 Hive Diver 完成的。
- 所有的数据都存储在 Hadoop 的 HDFS 中。
- 查询计划被转换为 MapReduce 任务，在 Hadoop 中执行（但要注意有些查询也可能没有 MapReduce 任务，如 select * from table）。
- Hadoop 和 Hive 都是用 UTF-8 编码的。

总之，Hive 是 Hadoop 的延伸，是一个提供了查询功能的数据仓库核心组件，Hadoop 底层的 HDFS 为 Hive 提供了数据存储，MapReduce 为 Hive 提供了分布式运算。HDFS 上存储着海量的数据，如果要对这些数据进行计算和分析，那么需要使用 Java 编写 MapReduce 程序来实现，但 Java 编程门槛较高，且一个 MapReduce 程序写起来要几十、上百行。而 Hive 可以直接通过 SQL 操作 Hadoop，SQL 简单易写、可读性强，Hive 将用户提交的 SQL 解析成 MapReduce 任务供 Hadoop 直接运行。Hive 某种程度来说也不进行数据计算，只是个解释器，只负责将用户对数据处理的逻辑通过 SQL 编程提交后解释成 MapReduce 程序，然后将这个 MapReduce 程序提交给 YARN 进行调度执行，因此，实际进行分布式运算的是 MapReduce 程序。

1.5 Hive 和关系数据库的异同

Hive 数据仓库与传统意义上的数据库是有区别的。一般来说，基于传统方式，可以用 Oracle 数据库或 MySQL 数据库来搭建数据仓库，数据仓库中的数据保存在 Oracle 或 MySQL 数据库中。Hive 数据仓库和它们不同的是，Hive 数据仓库建立在 Hadoop 集群的 HDFS 之上，也就是说，Hive 数据仓库中的数据是保存在 HDFS 上的。Hive 数据仓库可以通过 ETL 的形式来抽取、转换和加载数据。Hive 提供了类似 SQL 的查询语句 HQL，可以用"select*from 表名;"来查询 Hive 数据仓库中的数据，这与关系数据库的操作是一样的。

关系数据库都是为实时查询的业务而设计的，而 Hive 则是为对海量数据进行数据挖掘而设计的，Hive 实时性差，但它很容易扩展自己的存储能力和计算能力。Hive 与关系数据库的对比如表 1-1 所示。

表1-1 Hive与关系数据库的对比

对 比 项	Hive	RDBMS
查询语言	HQL	SQL
数据存储	HDFS	Raw Device 或 Local FS
数据格式	没有定义专门的数据格式	有专门的数据格式
数据更新	不支持对数据的改写和添加	允许添加、修改数据
索引	无	有
执行	MapReduce	Executor
执行延迟	高	低
可扩展性	可扩展性与 Hadoop 一致	受 ACID 语义的严格限制、扩展性非常有限
处理数据规模	大	小

（1）查询语言。由于 SQL 被广泛地应用在数据仓库中，因此，专门针对 Hive 的特性设计了类 SQL 的查询语言 HQL。熟悉 SQL 开发的开发者可以很方便地使用 Hive 进行开发。

（2）数据存储位置。Hive 是建立在 Hadoop 之上的，所有 Hive 的数据都存储在 HDFS 中。数据库则可以将数据保存在块设备或者本地文件系统中。

（3）数据格式。Hive 中没有定义专门的数据格式，数据格式可以由用户指定，用户定义数据格式需要指定三个属性：列分隔符（通常为空格、\t、\x001）、行分隔符（\n）以及读取文件数据的方法（Hive 中默认有三个文件格式，即 TextFile、SequenceFile 以及 RCFile）。由于在加载数据的过程中，不需要进行从用户数据格式到 Hive 定义的数据格式的转换，因此，Hive 在加载的过程中不会对数据本身进行任何修改，而只是将数据内容复制或者移动到相应的 HDFS 目录中。在数据库中，不同的数据库有不同的存储引擎，都定义了自己的数据格式，并且所有数据都会按照一定的组织形式进行存储，因此，数据库加载数据的过程会比较耗时。

（4）数据更新。由于 Hive 是针对数据仓库应用设计的，而数据仓库的内容读多写少，因此，

Hive 不支持对数据的改写和添加，所有的数据都是在加载的时候就确定好的。数据库中的数据通常需要进行修改，因此可以使用 INSERT INTO...VALUES 添加数据，使用 UPDATE...SET 修改数据。

（5）索引。之前已经说过，Hive 在加载数据的过程中不会对数据进行任何处理，甚至不会对数据进行扫描，因此也没有对数据中的某些 key 建立索引。Hive 要访问数据中满足条件的特定值时，需要暴力扫描整个数据，因此访问延迟较高。由于 MapReduce 的引入，Hive 可以并行访问数据，因此即使没有索引，对于大数据量的访问，Hive 仍然可以体现出优势。数据库中，通常会针对一个或者几个列建立索引，因此对于少量的特定条件的数据的访问，数据库可以有很高的效率、较低的延迟。由于数据的访问延迟较高，因此决定了 Hive 不适合在线数据查询。

（6）执行。Hive 中大多数查询的执行是通过 Hadoop 提供的 MapReduce 来实现的（类似 select * from tbl 的查询不需要 MapReduce）。而数据库通常有自己的执行引擎。

（7）执行延迟。之前提到，Hive 在查询数据的时候，由于没有索引，需要扫描整个表，因此延迟较高。另外一个导致 Hive 执行延迟高的因素是 MapReduce 框架。由于 MapReduce 本身具有较高的延迟，因此在利用 MapReduce 执行 Hive 查询时，也会有较高的延迟。相对地，数据库的执行延迟较低。当然，这个低是有条件的，即数据规模较小，当数据规模大到超过数据库的处理能力的时候，Hive 的并行计算显然更有优势。

（8）可扩展性。由于 Hive 建立在 Hadoop 之上，因此 Hive 的可扩展性和 Hadoop 的可扩展性是一致的（国内规模较大的 Hadoop 集群平台提供者有百度和阿里巴巴等大型互联网企业。截至目前，百度 Hadoop 集群规模达到近十个，单集群超过 2800 台机器节点，Hadoop 机器总数有上万台）。而数据库由于 ACID 语义的严格限制，扩展性非常有限。目前最先进的并行数据库 Oracle 在理论上的扩展能力也只有 100 台左右。

（9）数据规模。由于 Hive 建立在集群上并可以利用 MapReduce 进行并行计算，因此可以支持很大规模的数据。数据库可以支持的数据规模较小。

1.6 Hive 数据存储简介

首先，Hive 没有专门的数据存储格式，也没有为数据建立索引，用户可以非常自由地组织 Hive 中的表，只需要在创建表的时候告诉 Hive 数据中的列分隔符和行分隔符，Hive 就可以解析数据。

其次，Hive 中所有的数据都存储在 HDFS 中。Hive 中包含以下数据模型：Table、External Table、Partition、Bucket。

1）Hive

Hive 中的 Table 和数据库中的 Table 在概念上是类似的，每一个 Table 在 Hive 中都有一个相应的目录存储数据。例如，一张表 htduan，它在 HDFS 中的路径为/warehouse/htduan，其中，warehouse 是在 hive-site.xml 中由${hive.metastore.warehouse.dir}指定的数据仓库的目录，所有的 Table 数据（不包括 External Table）都保存在这个目录中。

2）External Table

External Table 指向已经存储在 HDFS 中的数据，可以创建 Partition。它和 Table 在元数据的组织上是相同的，而实际数据的存储则有较大的差异。

对于 Table 的创建过程和数据加载过程（这两个过程可以在同一个语句中完成），在加载数据的过程中，实际数据会被移动到数据仓库目录中，之后对数据的访问将会直接在数据仓库目录中完成；删除表时，表中的数据和元数据将会被同时删除。External Table 只有一个过程，加载数据和创建表同时完成（CREATE EXTERNAL TABLE...LOCATION），实际数据是存储在 LOCATION 后面指定的 HDFS 路径中，并不会移动到数据仓库目录中；当删除一个 External Table 时，仅删除了 Hive 的元数据。

3）Partition

Partition 对应于数据库中的 Partition 列的密集索引，但是 Hive 中 Partition 的组织方式和数据库中的有很大不同。在 Hive 中，表中的一个 Partition 对应于表下的一个目录，所有的 Partition 的数据都存储在对应的目录中。例如，htduan 表中包含 dt 和 ctry 两个 Partition，则对应于 dt=20100801、ctry=US 的 HDFS 子目录为/warehouse/htduan/dt=20100801/ctry=US，对应于 dt=20100801、ctry=CA 的 HDFS 子目录为/warehouse/htduan/dt=20100801/ctry=CA。

4）Bucket

Bucket 对指定列计算 hash，根据 hash 值切分数据，目的是并行。每一个 Bucket 对应一个文件。例如，将 user 列分散至 32 个 Bucket，对 user 列的值计算 hash，对应 hash 值为 0 的 HDFS 目录为/warehouse/htduan/dt=20100801/ctry=US/part-00000，hash 值为 20 的 HDFS 目录为/warehouse/htduan/dt=20100801/ctry=US/part-00020。

第2章

Hive部署与基本操作

2.1 Linux环境的搭建

2.1.1 VirtualBox虚拟机安装

本书将使用VirtualBox作为虚拟环境安装Linux和Hadoop。VirtualBox最早由SUN公司开发。由于SUN公司目前已经被Oracle收购，因此可以在Oracle公司的官方网站上下载VirtualBox虚拟机软件的安装程序，产品地址为https://www.virtualbox.org。笔者写作本书时，VirtualBox的最新版本为7.0.6。

首先，到VritualBox的官方网站下载Windows hosts版本的VirtualBox。下载页面地址为https://www.virtualbox.org/wiki/Downloads，页面如图2-1所示。

VirtualBox 7.0.6 platform packages

- Windows hosts
- macOS / Intel hosts
- Developer preview for macOS / Arm64 (M1/M2) hosts
- Linux distributions
- Solaris hosts
- Solaris 11 IPS hosts

图2-1　VirtualBox下载地址

同时，VitualBox需要虚拟化CPU的支持，如果安装的操作系统是不支持x64位的CentOS，那么可以在宿主机开机时按F12键进入宿主机的BIOS设置界面，并打开CPU的虚拟化设置界面。CPU的虚拟化设置界面如图2-2所示。

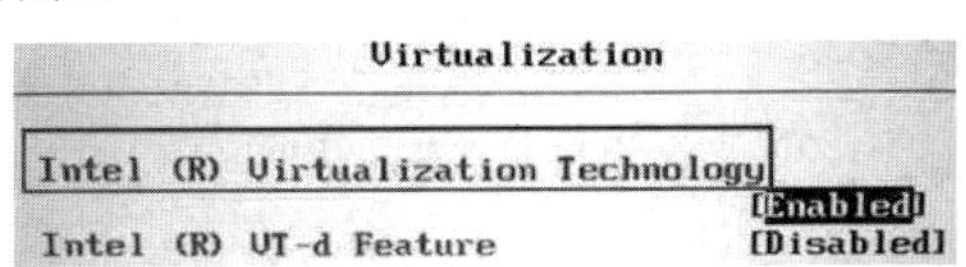

图2-2　CPU的虚拟化设置

读者下载完成 VirtualBox 虚拟机后，自行安装即可。虚拟机的安装相对比较简单，以下是重要安装环节的截图。

网络功能的安装界面如图 2-3 所示，在该界面单击“是”按钮。

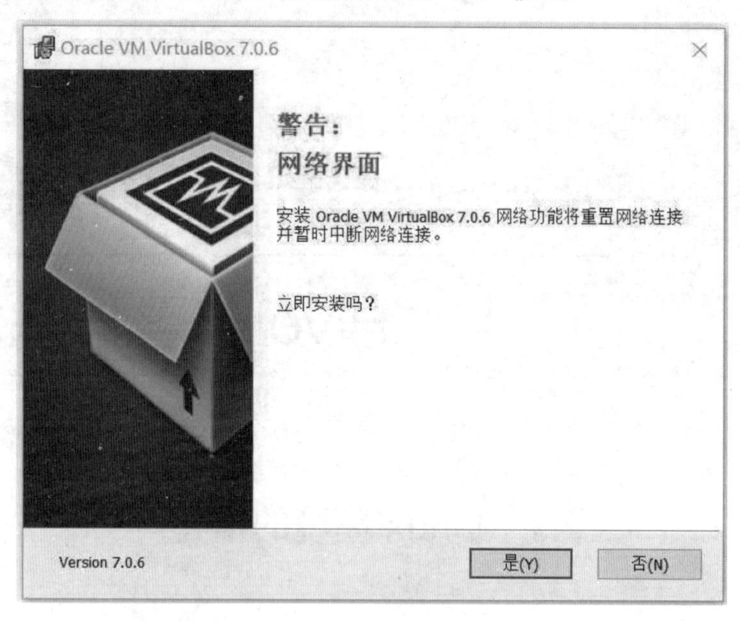

图 2-3　网络功能安装

网络功能下一步的安装界面如图 2-4 所示，在该界面单击“安装”按钮。

图 2-4　网络功能安装

网络功能安装成功后，会在“网络连接”里面多出一个名为 Virtual Box Host Only 的本地网卡，此网卡用于宿主机与虚拟机通信，如图 2-5 所示。

图 2-5　本地虚拟网卡

2.1.2 安装 Linux 操作系统

本书将使用 CentOS 7 作为操作系统环境来学习和安装 Hadoop。首先需要下载 CentOS 操作系统，下载 Minimal（最小）版本的即可，因为我们使用的 CentOS 并不需要可视化界面。

CentOS 的官方网址为 https://www.centos.org/。CentOS 7 的下载页面如图 2-6 所示。

CentOS-7-x86_64-DVD-2009.iso	4.4 GiB	2020-11-04 19:37
CentOS-7-x86_64-DVD-2009.torrent	176.1 KiB	2020-11-06 22:44
CentOS-7-x86_64-Everything-2009.iso	9.5 GiB	2020-11-02 23:18
CentOS-7-x86_64-Everything-2009.torrent	380.6 KiB	2020-11-06 22:44
CentOS-7-x86_64-Minimal-2009.iso	973.0 MiB	2020-11-03 22:55
CentOS-7-x86_64-Minimal-2009.torrent	38.6 KiB	2020-11-06 22:44

图 2-6　CentOS 下载页面

下载完成以后，将得到一个 CentOS-7-x86_64-Minimal-2009.iso 文件。注意文件名中的 2009 不是指 2009 年，而是指 2020 年 09 月发布的版本。接下来，启动 VirtualBox，操作步骤如下：

步骤 01 在 VirtualBox 主界面的菜单栏上单击“新建”按钮，如图 2-7 所示。

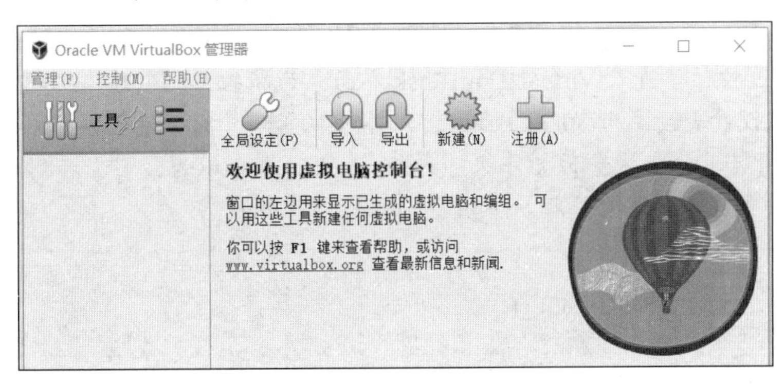

图 2-7　VirtualBox 启动界面

步骤 02 在 Virtual machine Name and Operating System 界面设置虚拟机的名称为 CentOS7-201，保持 Folder 为默认值，选择操作系统镜像，如图 2-8 所示。

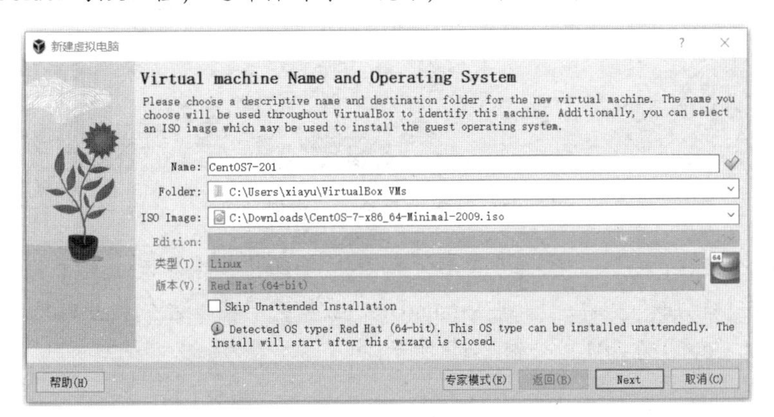

图 2-8　选择将要安装的操作系统

步骤03 单击 Next 按钮，进入 Hardware 界面，为新的系统分配内存，建议设置为 4GB（最少 2GB）或以上，具体根据宿主机的内存而定；同时建议设置 CPU 为 2 颗，如图 2-9 所示。

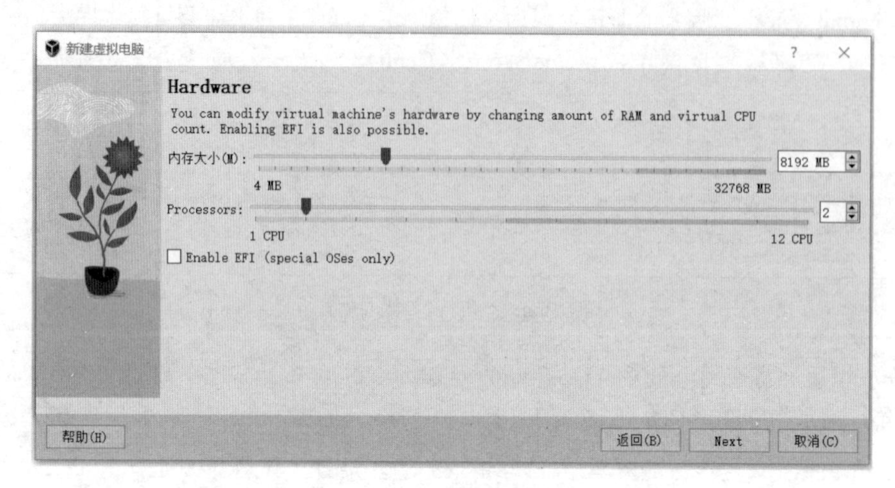

图 2-9 为新的系统分配内存

步骤04 单击 Next 按钮，进入 Virtual Hard disk 界面，为新的系统创建虚拟硬盘，设置为动态增加，建议最大设置为 30GB 或以上，如图 2-10 所示。同时设置虚拟文件的保存目录，默认的情况下，会将虚拟文件保存到 C:\盘上。笔者以为最好保存到非系统盘上，如 D:\OS 目录将是不错的选择。

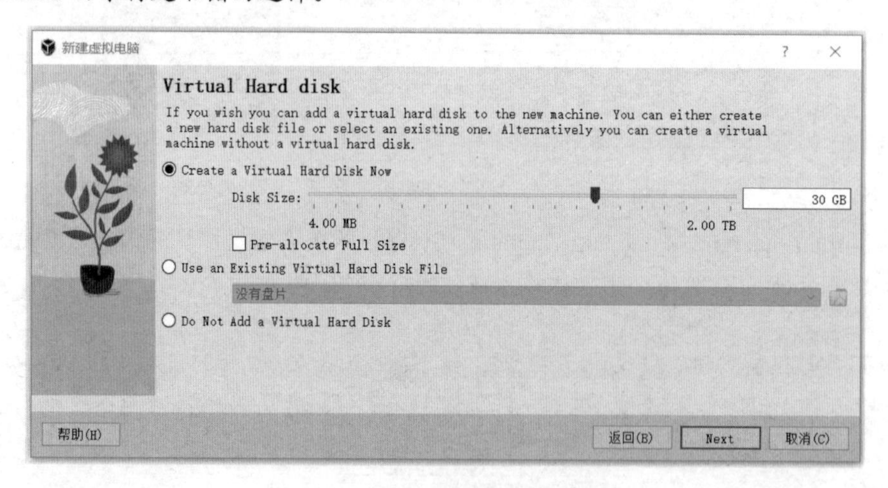

图 2-10 为新的系统创建虚拟硬盘

步骤05 单击 Next 按钮，进入“摘要”界面，如图 2-11 所示。在界面上单击 Finish 按钮，关闭“新建虚拟电脑”窗口，回到 VirsualBox 主窗口，窗口左侧栏此时已经显示我们新建的虚拟机 CentOS7-201，如图 2-12 所示。

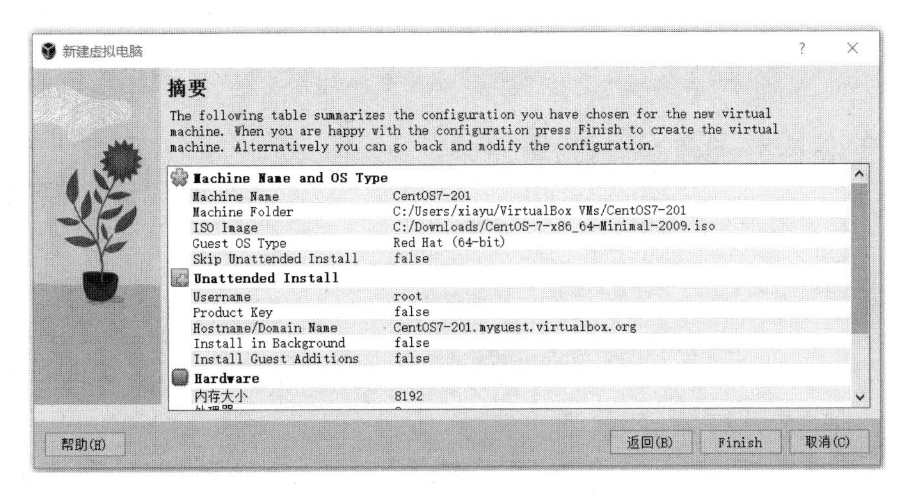

图 2-11　“摘要”界面

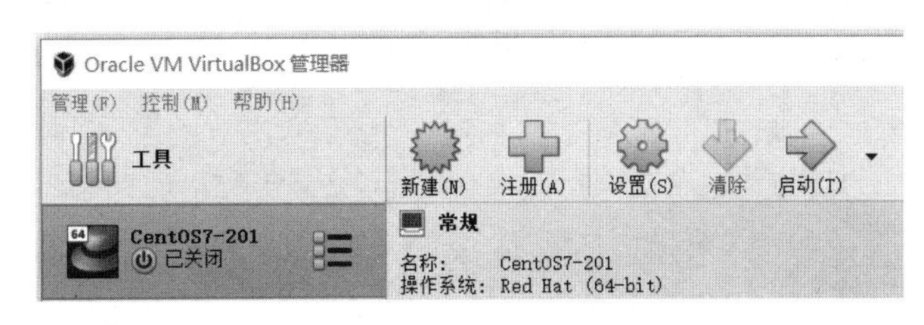

图 2-12　显示新建的虚拟机

步骤 06 在图 2-12 所示的 VirsualBox 主窗口左侧选中 CentOS7-201 虚拟机，并单击右上方的“设置”按钮，打开“CentOS7-201-设置”窗口，如图 2-13 所示。

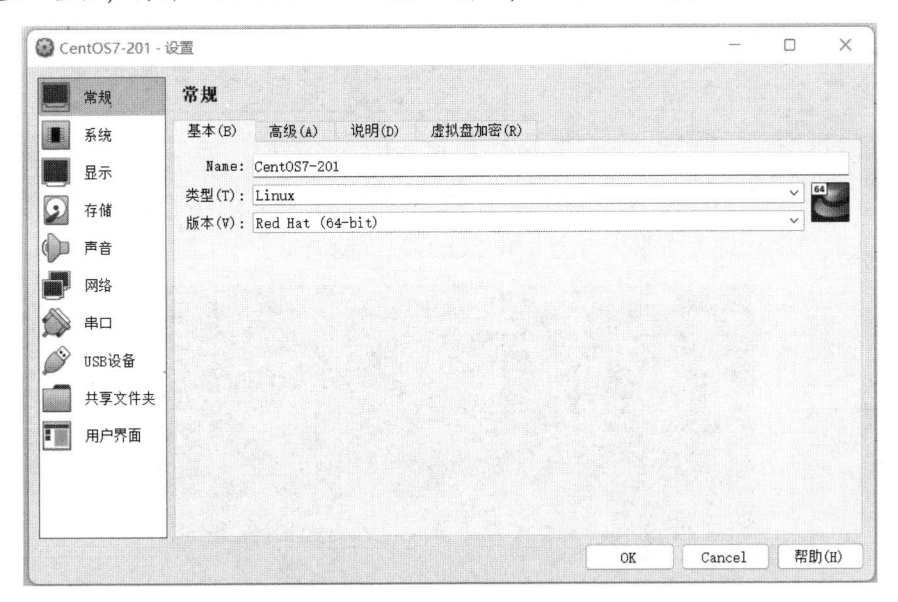

图 2-13　“CentOS7-201-设置”窗口

步骤 07 在“CentOS7-201-设置”窗口左侧选择“网络”，右侧窗体会显示“网络”设置界面，将网卡 1 的连接方式设置为 NAT 用于连接外网，将网卡 2 的连接方式设置为 Host-Only

用于与宿主机进行通信。如果没有网卡 2，则需要关闭 Linux 虚拟机，在这个设置界面上对网卡 2 进行“启用网络连接”设置，并选择连接方式为“仅主机(Host-Only)网络”。

网卡 1 的设置如图 2-14 所示。

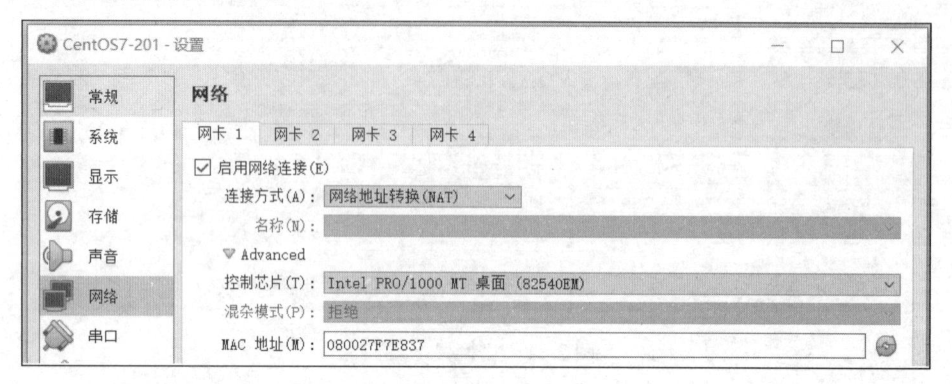

图 2-14 网卡 1 的设置

网卡 2 的设置如图 2-15 所示。

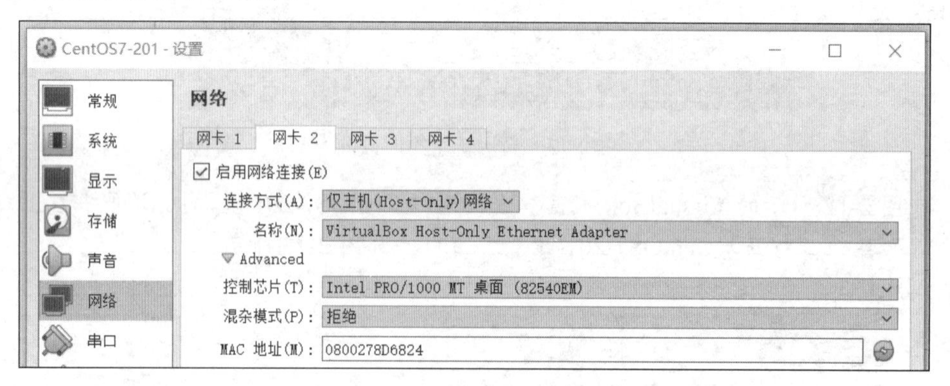

图 2-15 网卡 2 的设置

步骤 08 启动 CentOS7-201 虚拟机，进入安装 CentOS 7 的界面，选择 Install CentOS Linux 7，如图 2-16 所示。接下来就开始安装 CentOS Linux 了。

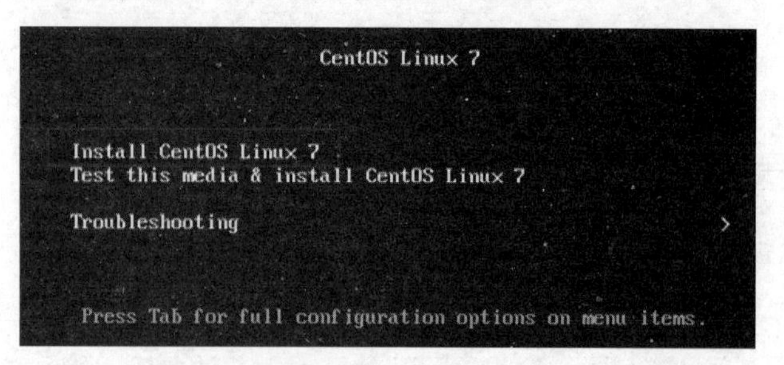

图 2-16 选择 Install CentOS Linux 7

步骤 09 在安装过程中出现选择语言项目，可以选择“中文”。单击如图 2-17 所示的界面来选择安装介质，同时进入安装位置，选择整个磁盘即可，选择磁盘后的界面如图 2-18 所示。注意，必须同时选择打开 CentOS 的网络，如图 2-19 和图 2-20 所示，否则安装成功以后，CentOS 将没有网卡设置的选项。

图 2-17 选择安装介质

图 2-18 选择磁盘

图 2-19 打开 CentOS 网络 1

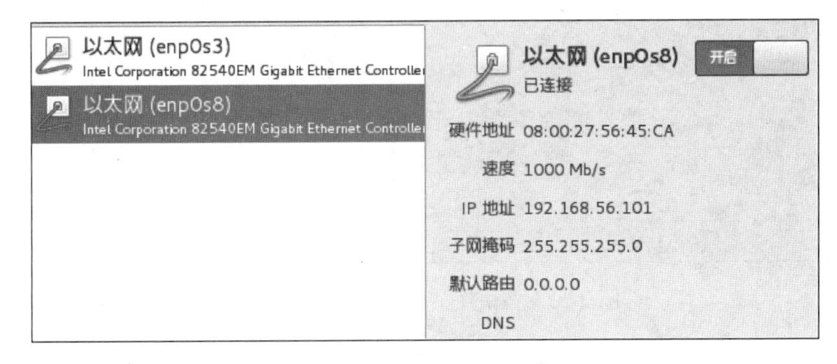

图 2-20 打开 CentOS 网络 2

步骤 10 在安装过程中，创建一个非 root 用户，并选择属于管理员组（输入的密码务必牢记），如图 2-21 和图 2-22 所示。在其后的操作中，笔者不建议使用 root 账户进行具体的操作。一般情况下，使用这个非 root 用户执行 sudo 命令即可用 root 账户执行相关命令

图 2-21 创建非 root 用户

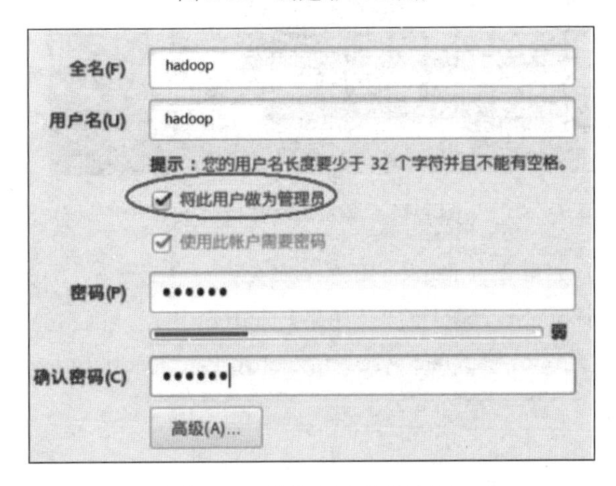

图 2-22 设置用户为管理员

步骤 11 在安装完成以后，重新启动并测试是否可以使用刚创建的用户名和密码登录。刚开始

安装完成后，右击虚拟机，在弹出的快捷菜单中选择“启动”→“正常启动”（即以有界面的方式启动），如图2-23所示。等我们设置好一些信息后，即可以选择“无界面启动”。

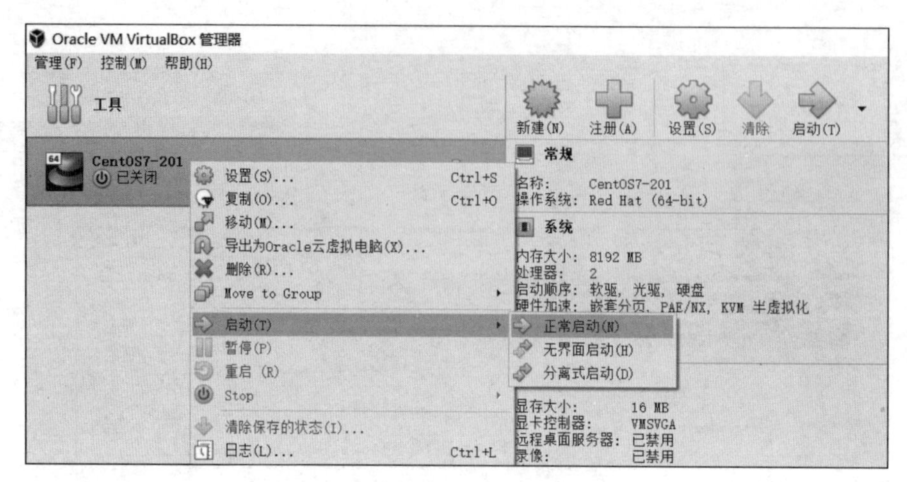

图2-23 在快捷菜单中选择正常启动

步骤12 设置静态IP地址。启动后，将显示如图2-24所示的界面，此时可以选择以root用户名和密码登录。注意，输入密码时将不会有任何的响应，不用担心，只要确认输入正确，按Enter键后即可看到登录成功后的界面，如图2-25所示。

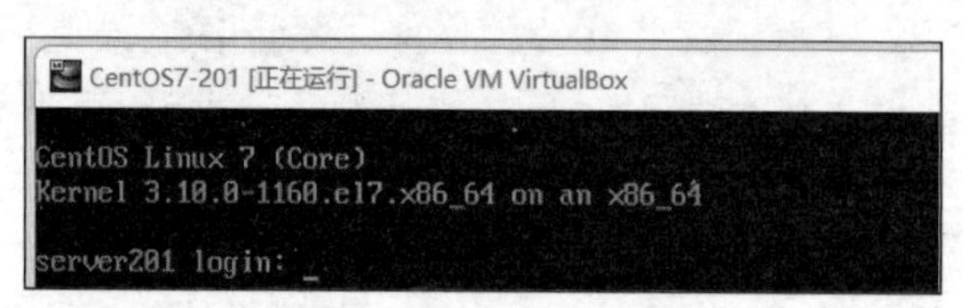

图2-24 登录界面

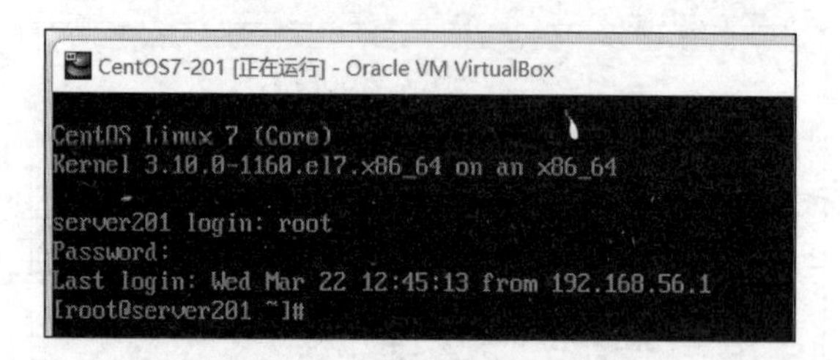

图2-25 登录成功后的界面

对于Linux系统来说，如果当前是root用户，那么将会显示“#”，如图2-25所示，root用户登录成功后，将会显示“[root@server8~]#”。如果是非root用户，则将显示为“$”。

设置静态IP地址，使用vim修改/etc/sysconfig/network-scripts/ifcfg-enp0s8，修改内容如下：

```
TYPE=Ethernet
PROXY_METHOD=none
BROWSER_ONLY=no
BOOTPROTO=static
DEFROUTE=yes
```

```
IPV4_FAILURE_FATAL=no
IPV6_INIT=yes
IPV6_AUTOCONF=yes
IPV6_DEFROUTE=yes
IPV6_FAILURE_FATAL=no
IPV6_ADDR_GEN_MODE=stable-privacy
NAME=enp0s8
UUID=620377da-1744-4268-b6d6-a519d27e01c6
DEVICE=enp0s8
ONBOOT=yes
IPADDR=192.168.56.201
```

其中 IPADDR=192.168.56.201 为本 Linux 的 Host-Only 网卡地址，用于主机通信。输出完成以后，按 ESC 键，然后再输入“:wq”保存退出即可。这是 vim 的基本操作，不了解的读者可以去网上查看一下 vim 的基本使用方法。

务必牢记上面设置的 IP 地址，这个 192.168.56.201 的 IP 地址在后面会经常出现。现在可以关闭系统，并以非界面方式重新启动 CentOS。以后我们将使用 SSH 客户端登录此 CentOS。

ifcfg-enp0s8 文件是在配置了 Host-Only 网卡的情况下才会存在，如果没有这个文件，则关闭 Linux，并重新添加 Host-Only 网卡，再行配置。如果添加了 Host-Only 网卡后依然没有此文件，那么可以在相同目录下复制 ifcfg-enp0s3 为 ifcfg-enp0s8 后，再进行配置。

现在关闭 CentOS，以无界面方式启动，如图 2-26 所示。

图 2-26　以无界面方式启动

注意：

（1）本书重点不是讲 VirtualBox 虚拟机的使用，因此这里只给出关键的操作步骤。

（2）在安装过程中，鼠标会在虚拟机和宿主机之间切换。如果要从虚拟机中退出鼠标，按键盘右边的 Ctrl 键即可。

（3）登录 Linux 系统后，随手执行命令“yum-y install vim”安装上 vim，方便使用。

（4）对于 Linux 命令，读者可自行参考 Linux 手册，如 vim/vi、sudo、ls、cp、mv、tar、chmod、chown、scp、ssh-keygen、ssh-copy-id、cat、mkdir 等，将是后面经常使用的命令。

2.1.3　SSH 工具与使用

Linux 安装成功后，系统自动运行 SSH 服务，读者可以选择 XShell、CRT、MobaXterm 等客户端作为 Linux 远程命令行操作工具，同时配合它们的 xFtp 可以实现文件的上传与下载。XShell 和 CRT 是收费软件，不过读者在安装时选择 free for school（学校免费版本）即可免费使用。

MobaXterm 个人版是免费的，本书选用它作为远程命令行执行、文件上传下载以及配置文件编辑的工具。到官网下载 MobaXterm 并安装完成以后，配置一下 SSH 即可登录 Linux 系统。配置很简单，操作步骤如下：

步骤 01 在 MobaXterm 主界面上单击左上方的 Session 按钮，如图 2-27 所示。

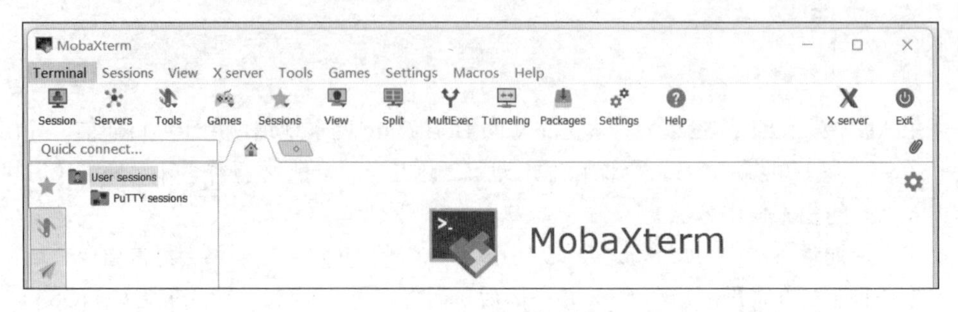

图 2-27　创建新的连接

步骤 02 单击 Session 按钮后弹出如图 2-28 所示的窗口，在窗口上单击 SSH 按钮，在相应的文本框中输入主机名称和登录用户名，再单击窗口下方的 OK 按钮保存一下。

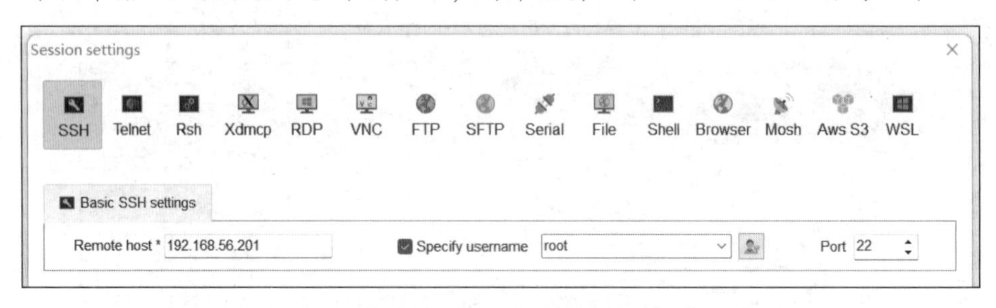

图 2-28　输入主机名称并登录

步骤 03 此时会打开 Linux 交互界面，提示输入 root 密码，输入密码不会有任何的回显，只要输入正确，按 Enter 键即可登录，如图 2-29 所示。

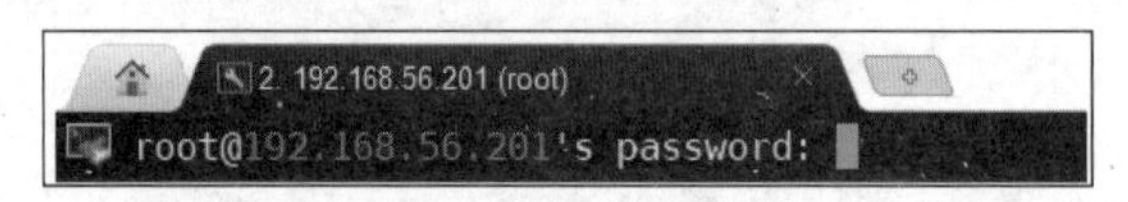

图 2-29　输入密码

root 用户登录成功以后的界面如图 2-30 所示，用户可以通过这个界面操作 Linux 系统。

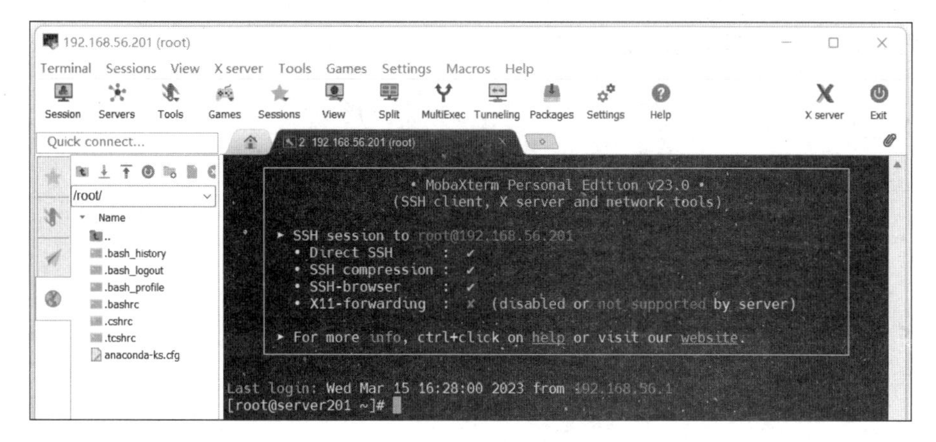

图 2-30　SSH 登录成功

还可以配置 SFTP 连接，方便本地下载的 Linux 软件包上传到 Linux 系统进行安装配置，Linux 系统上的配置文件也可以在本地编辑后自动上传。SFTP 登录界面如图 2-31 所示。

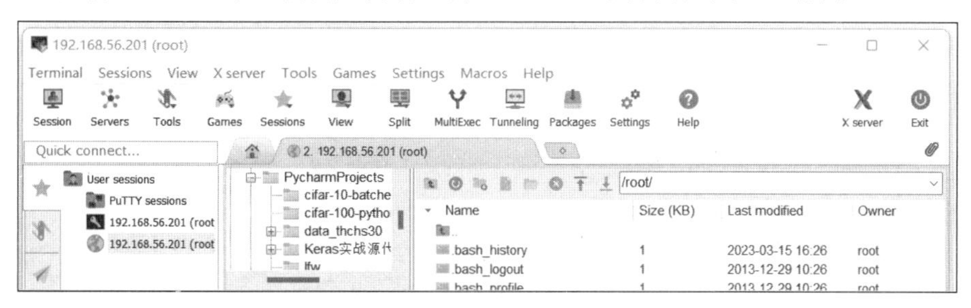

图 2-31　MobaXterm 文件上传

2.1.4　Linux 统一设置

后面配置 Hadoop 环境时将使用一些 Linux 统一的设置，在此一并列出。由于本次登录（如图 2-28 所示）是用 root 登录的，因此可以直接操作某些命令，不用添加 sudo 命令。

1. 配置主机名称

笔者习惯将“server+IP 最后一部分数字”作为主机名称，所以取主机名为“server201”，因为本主机设置的 IP 地址是 192.168.56.201。

```
# hostnamectl set-hostname server201
```

2. 修改 hosts 文件

通过 vim /etc/hosts 命令在 hosts 文件的最后添加以下配置：

```
192.168.56.201    server201
```

3. 关闭且禁用防火墙

```
# systemctl stop firewalld
# systemctl disable firewalld
```

4. 禁用 SELinux，需要重新启动

```
#vim /etc/selinux/config
SELINUX=disabled
```

5. 设置时间同步（可选）

```
#vim /etc/chrony.conf
```

删除所有的 server 配置，只添加：

```
server ntp1.aliyun.com iburst
```

重新启动 chronyd：

```
#systemctl restart chronyd
```

查看状态：

```
#chronyc sources -v
^* 120.25.115.20
```

如果结果显示“*”，则表示时间同步成功。

6. 在/usr/java 目录下安装 JDK1.8

usr 目录的意思是 unix system resource 目录，可以将 JDK1.8_x64 安装到此目录下。首先去 Oracle 网站下载 JDK1.8 的 Linux 版本，页面如图 2-32 所示。

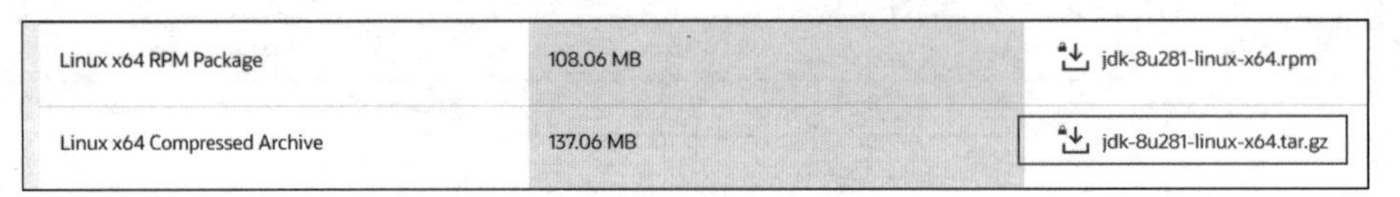

图 2-32　JDK 下载

然后将安装包上传到 Linux 并解压：

```
# mkdir /usr/java
# tar -zxvf jdk-8u281-linux-x64.tar.gz -C /usr/java/
```

7. 配置 JAVA_HOME 环境变量

```
# vim /etc/profile
```

在 profile 文件最后，添加以下配置：

```
export JAVA_HOME=/usr/java/jdk1.8.0_281
export PATH=.:$PATH:$JAVA_HOME/bin
```

让环境变量生效：

```
# source /etc/profile
```

检查 Java 版本：

```
[root@localhost bin]# java -version
java version "1.8.0_281"
Java(TM) SE Runtime Environment (build 1.8.0_192-b12)
```

```
Java HotSpot(TM) 64-Bit Server VM (build 25.192-b12, mixed mode)
```

至此，基本的 Linux 运行环境就已经配置完成了。

提示：在 VirtualBox 虚拟机中，可以通过复制的方式快速创建一个虚拟机，因为本小节已经为统一设置的 CentOS 镜像文件创建了副本，用于备份或者搭建集群。

2.2　Hadoop 伪分布式环境的搭建

因为 Hive 运行在 Java 和 Hadoop 之上，所以在运行 Hive 之前，首先确保已经安装好 Java 和 Hadoop，并启动 Hadoop。本节重点介绍 Hadoop 伪分布式环境的搭建，读者如果对 Hadoop 本地模式比较熟悉，那么可以跳过 2.2.1 节，直接搭建 Hadoop 伪分布式环境，本书讲解的 Hive 主要运行在这个环境下。

2.2.1　安装本地模式运行的 Hadoop

本地模式运行的 Hadoop 可以帮助我们快速运行一个 MapReduce 示例，以了解其运行方法。

1. 下载 Hadoop

Hadoop3.2.3 的下载地址为：

```
https://www.apache.org/dyn/closer.cgi/hadoop/common/hadoop-3.2.3/hadoop-3.2
.3.tar.gz
```

2. 解压并配置环境

以 hadoop 用户登录，并在/home/hadoop 的主目录下创建一个目录，用于安装 Hadoop：

```
$ mkdir ~/program
```

上传 Hadoop 压缩包，并解压到 program 目录下：

```
$ tar -zxvf hadoop-3.2.3.tar.gz -C ~/program/
```

配置 Java 的环境变量，修改 hadoop 解压目录下的/etc/hadoop/hadoop-env.sh 文件，找到${JAVA_HOME}并设置为本机 JAVA_HOME 的地址。

```
$ vim ~/program/hadoop-3.2.3/etc/hadoop/hadoop-env.sh
export JAVA_HOME=/usr/java/jdk1.8.0_281
```

配置 hadoop 用户的环境变量：

```
$ vim /home/hadoop/.bash_profile
export HADOOP_HOME=/home/hadoop/program/hadoop-3.2.3
export PATH=$PATH:$HADOOP_HOME/bin
```

注意：由于笔者是用 hadoop 用户登录的，因此只配置了 hadoop 用户的环境变量，这种情况下，这种配置只会让当前用户可用。读者可以根据自己的要求进行配置，比如：如果配置到/etc/profile 文件中，则是整个系统都可以使用的环境变量，这种情况下，就不要将 hadoop 安装到某个用户的主目录下了。

让环境变量生效：

```
$source ~/.bash_profile
```

输入 hadoop version 命令，查看 Hadoop 的版本：

```
[hadoop@server201 ~]$ hadoop version
Hadoop 3.2.3
Source code repository Unknown -r 7a3bc90b05f257c8ace2f76d74264906f0f7a932
Compiled by hexiaoqiao on 2021-01-03T09:26Z
Compiled with protoc 2.5.0
From source with checksum 5a8f564f46624254b27f6a33126ff4
This command was run using
/home/hadoop/program/hadoop-3.2.3/share/hadoop/common/hadoop-common-3.2.3.jar
```

3. 独立运行 MapReduce

Hadoop 可以运行在一个非分布式的环境下，即可以运行为一个独立的 Java 进程。现在运行一个 WordCount 的 MapReduce 示例。

首先创建一个文本文件 a.txt，并输入几行英文句子：

```
[hadoop@server201 ~]$ touch a.txt
[hadoop@server201 ~]$ vim a.txt
Hello This is
a Very Sample MapReduce
Example of Word Count
Hope You Run This Program Success!
```

然后进行 WordCount 测试：

```
[hadoop@server201 ~]$ hadoop jar \
~/program/hadoop-3.2.3/share/hadoop/mapreduce/hadoop-mapreduce-examples-3.2
.3.jar \
 wordcount \
~/a.txt \
~/out
```

命令说明：

- hadoop jar 用于执行一个 MapReduce 示例。在 Linux 中，如果命令有多行，那么可以通过输入\（斜线）来换行。注意\前面必须有空格。
- hadoop-mapreduce-examples-3.2.3.jar 为官网提供的示例程序包，wordcount 是执行的任务，~/a.txt 是输入的目录或文件，~/out 是程序执行成功以后的输出目录。

命令执行成功后，会显示以下信息（注意，输出的日志会比较多，请仔细查找）：

```
2021-03-08 21:59:19,536 INFO mapreduce.Job:  map 100% reduce 100%
2021-03-08 21:59:19,537 INFO mapreduce.Job: Job job_local215774179_0001
completed successfully
```

程序执行成功以后，进入 out 输出目录，查看输出目录中的数据文件：

```
[hadoop@server201 ~]$ cd out/
```

```
[hadoop@server201 out]$ ll
总用量 4
-rw-r--r-- 1 hadoop hadoop 122 3月   8 21:59 part-r-00000
-rw-r--r-- 1 hadoop hadoop   0 3月   8 21:59 _SUCCESS
```

其中 part-r-0000 为数据文件；_SUCCESS 为标识成功的文件，里面没有数据。通过 cat 命令查看 part-r-00000 文件中的数据，可以看到已经对 a.txt 中的单词进行了数量统计，且默认排序为字母的顺序，字母后面跟的是此单词出现的次数。

```
[hadoop@server201 out]$ cat *
Count       1
Example     1
Hello       1
Hope        1
MapReduce   1
Program     1
Run         1
Sample      1
Success!    1
This        2
Very        1
Word        1
You         1
a           1
is          1
of          1
```

至此，独立运行模式的 Hadoop 已经可以成功运行。

Hadoop 独立运行方式只是一个练习，在正式的生产环境中，不会使用这种方式，这里只是让大家了解一下 MapReduce 的运行，而且在此模式下，Hadoop 的 HDFS 不会运行，也不会存储数据。

注意：下一小节继续讲解 Hadoop 伪分布式环境的搭建，本书有关 Hive 的安装配置和基本命令测试将会运行在 Hadoop 伪分布式环境下，因此读者务必要掌握这个环境的安装、配置与运行。

2.2.2 Hadoop 伪分布式环境的准备

注意：读者可以从 2.1 节配置好的 CentOS 虚拟机中复制一份干净的系统出来，用于本小节搭建 Hadoop 伪分布式环境，复制出来的 CentOS 虚拟机名称为 CenOS7-201。

Hadoop 伪分布式也是在单机模式下运行 Hadoop，但用不同的 Java 进程模仿分布式运行中的各类节点。这种模式下，我们需要运行 5 个守护进程，即 3 个负责 HDFS 存储的进程和 2 个负责 MapReduce 计算的进程。

负责 HDFS 存储的 3 个进程如图 2-33 所示。

- NameNode 进程作为主节点，主要负责分配数据存储的具体位置。
- SecondaryNameNode 进程作为 NameNode 日志备份和恢复进程，用于避免数据丢失。
- DataNode 进程作为数据的存储节点，用于接收客户端的数据读写请求。

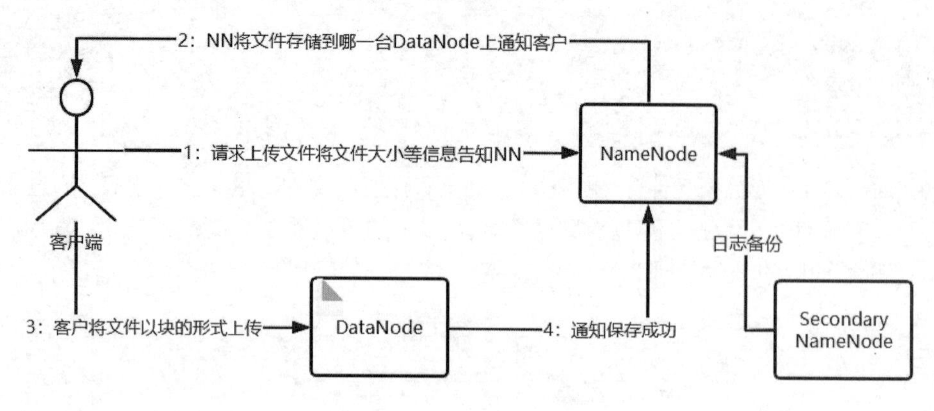

图 2-33 HDFS 守护进程

负责 MapReduce 计算的 2 个进程：

- ResourceManager 进程负责分配计算任务由哪一台主机执行。
- NodeManager 进程负责执行计算任务。

在真实集群环境下，这些进程部署的一般规则是：

（1）由于 NodeManger 需要读取 DataNode 上的数据用于执行计算，因此通常情况下 DataNode 与 NodeManger 并存。

（2）由于 NameNode 在运行时需要在内存中大量缓存文件块的数据，因此应该将它部署到内存比较大的主机上。

（3）在真实的集群环境下，一般部署多个 NameNode 节点，互为备份和切换关系，且不再部署 SecondaryNameNode 进程。

Hadoop 伪分布式可以让读者快速学习 HDFS 的命令及如何开发 MapReduce 应用，这对于学习 Hadoop 有很大的帮助。同时，本书讲解的 Hive 运行环境也主要配置在这个 Hadoop 伪分布式环境中。

在安装 Hadoop 伪分布式之前，笔者有以下建议：

- 配置静态 IP 地址。虽然是单机模式，但也建议配置静态的 IP 地址，这有助于以后配置集群环境时固定 IP，养成良好的习惯。
- 修改主机名为一个便于记忆的名称，如 server201，修改规则一般为将本机的 IP 地址的最后一段作为服务器的后缀，如 IP 地址为 192.168.56.201，则可以修改本主机的名称为 server201。
- 由于启动 Hadoop 的各个进程使用的是 SSH，因此必须配置本机免密码登录。本章后面的步骤会讲到如何配置 SSH 免密码登录。配置 SSH 免密码登录的规则是在启动的集群的主机上，向其他主机配置 SSH 免密登录，以便于操作机可以在不登录其他主机的情况下启动所需要的进程。
- 关闭防火墙。如果我们的 CentOS 7 没有安装防火墙，那么可以不用关闭。如果已经安装防火墙了，则检查防火墙的状态，如果是运行状态就关闭防火墙并禁用防火墙。注意，在生产环境下，不要直接禁用防火墙，而是指定 Hadoop 的某些端口开放。
- 使用非 root 用户，前面章节我们创建了一个名为 hadoop 的用户，此用户同时属于 wheel

组（拥有此组的用户可以使用 sudo 命令，执行一些 root 用户的操作），我们就将此用户作为执行命令的用户。

1. 配置静态 IP 地址

2.1.2 节已经讲解了静态 IP 地址的设置，此处再做一下补充。首先使用 SSH 登录 CentOS 7，然后使用 ifconfig 查看 IP 地址（如果没有 ifconfig 命令，那么可以使用 sudo yum -y install net-tools 安装 ifconfig 命令。其实在 CentOS 7 中，可以使用 ip addr 命令显示当前主机的 IP 地址，因此也可以不安装 net-tools）：

```
$ ifconfig
enp0s3: flags=4163<UP,BROADCAST,RUNNING,MULTICAST>  mtu 1500
inet 10.0.2.15  netmask 255.255.255.0  broadcast 10.0.2.255
enp0s8: flags=4163<UP,BROADCAST,RUNNING,MULTICAST>  mtu 1500
inet 192.168.56.201  netmask 255.255.255.0
```

结果显示为两块网卡，其中 enp0s3 的 IP 地址为 10.0.2.15，此网卡为 NAT 网络，可用于上网；enp0s8 的 IP 地址为 192.168.56.201，此网卡为 Host-Only 网络，用于与宿主机进行通信。我们要修改的就是 enp0s8 这个网卡，将它的 IP 地址设置为固定 IP。

IP 设置保存在/etc/sysconfig/network-scripts/ifcfg-enp0s8 文件中。使用 cd 命令，切换到这个目录下。使用 ls 显示这个目录下的所有文件，有可能只会发现 ifcfg-enp0s3 这个文件，可以使用 cp 命令将 ifcfg-enp0s3 复制一份为 ifcfg-enp0s8。由于 etc 目录不属于 hadoop 用户，因此在操作时需要添加 sudo 前缀。

```
$ sudo cp ifcfg-enp0s3 ifcfg-enp0s8
```

使用 vim 命令修改为静态 IP 地址：

```
$ sudo vim ifcfg-enp0s8
```

将原来的 dhcp 修改成 static，即静态的 IP 地址，并设置具体的 IP 地址。其中，每一个网卡都应该具有唯一的 UUID，因此建议修改任意的一个值，以便于与之前 enp0s3 的 UUID 不同。部分修改内容如下：

```
BOOTPROTO="static"
NAME="enp0s8"
UUID="d2a8bd92-cf0d-4471-8967-3c8aee78d101"
DEVICE="enp0s8"
IPADDR="192.168.56.201"
```

现在重新启动网络：

```
$ sudo systemctl restart network.service
```

重新启动网络后，再次查看：

```
[hadoop@server201 ~]$ ifconfig
enp0s3: flags=4163<UP,BROADCAST,RUNNING,MULTICAST>  mtu 1500
        inet 10.0.2.15  netmask 255.255.255.0  broadcast 10.0.2.255
enp0s8: flags=4163<UP,BROADCAST,RUNNING,MULTICAST>  mtu 1500
        inet 192.168.56.201  netmask 255.255.255.0  broadcast 0x20<link>
lo: flags=73<UP,LOOPBACK,RUNNING>  mtu 65536
```

```
        inet 127.0.0.1  netmask 255.0.0.0
```

可以发现 IP 地址已经发生了变化。

2. 修改主机名称

使用 hostname 命令检查当前主机的名称：

```
$ hostname
localhost
```

使用 hostnamectl 命令修改主机的名称：

```
$ sudo hostnamectl set-hostname server201
```

3. 配置 hosts 文件

hosts 文件是本地 DNS 解析文件。配置此文件，可以根据主机名找到对应的 IP 地址。

使用 vim 命令打开这个文件，并在文件中追加以下配置：

```
$ sudo vim /etc/hosts
192.168.56.201  server201
```

4. 关闭防火墙

默认情况下，CentOS 7 没有安装防火墙。我们可以通过命令 sudo firewall-cmd--state 检查防火墙的状态，如果显示 command not found，则表示没有安装防火墙，此步可以忽略。以下命令检查防火墙的状态：

```
$ sudo firewall-cmd --state
running
```

running 表示防火墙正在运行。以下命令用于停止和禁用防火墙：

```
$ sudo systemctl stop firewalld.service
$ sudo systemctl disable firewalld.service
```

5. 配置免密码登录

配置免密码登录的主要目的就是在使用 hadoop 脚本启动 Hadoop 的守护进程时，不需要再提示用户输入密码。SSH 免密码登录的主要实现机制就是在本地生成一个公钥，然后将公钥配置到需要被免密登录的主机上，登录时自己持有私钥与公钥进行匹配，如果匹配成功，则登录成功，否则登录失败。

可以使用 ssh-keygen 命令生成公钥和私钥文件，并将公钥文件复制到被 SSH 登录的主机上。以下是 ssh-keygen 命令，输入以后直接按两次 Enter 键，即可以生成公钥和私钥文件：

```
[hadoop@server201 ~]$ ssh-keygen -t rsa
Generating public/private rsa key pair.
Enter file in which to save the key (/home/hadoop/.ssh/id_rsa):
Created directory '/home/hadoop/.ssh'.
Enter passphrase (empty for no passphrase):
Enter same passphrase again:
Your identification has been saved in /home/hadoop/.ssh/id_rsa.
Your public key has been saved in /home/hadoop/.ssh/id_rsa.pub.
```

```
The key fingerprint is:
SHA256:IDI032gBEDXhFVE1l6oYca5P4fkfIZRywyhgJ4Id/I4 hadoop@server201
The key's randomart image is:
+---[RSA 2048]----+
|=*%+*+..o ..     |
|.=oO.+.o +.      |
|  +.*+= *.       |
|   +ooo=..       |
|   o = +S .      |
|  E + =  . .     |
|     o . .       |
|      . . .      |
|       ..        |
+----[SHA256]-----+
```

如上的提示信息所示，生成的公钥和私钥文件将被放到~/.ssh/目录下。其中 id_rsa 文件为私钥文件，rd_rsa.pub 为公钥文件。现在我们再使用 ssh-copy-id 命令将公钥文件发送到目标主机。由于是登录本机，因此直接输入本机的主机名即可：

```
[hadoop@server201 ~]$ ssh-copy-id server201
/usr/bin/ssh-copy-id: INFO: Source of key(s) to be installed:
"/home/hadoop/.ssh/id_rsa.pub"
The authenticity of host 'server201 (192.168.56.201)' can't be established.
ECDSA key fingerprint is SHA256:KqSRs/H1WxHrBF/tfM67PeiqqcRZuK4ooAr+xT5Z4OI.
ECDSA key fingerprint is MD5:05:04:dc:d4:ed:ed:68:1c:49:62:7f:1b:19:63:5d:8e.
Are you sure you want to continue connecting (yes/no)? yes 输入 yes
/usr/bin/ssh-copy-id: INFO: attempting to log in with the new key(s), to filter
out any that are already installed
/usr/bin/ssh-copy-id: INFO: 1 key(s) remain to be installed -- if you are prompted
now it is to install the new keys
```

输入密码后按 Enter 键，将会提示成功信息：

```
hadoop@server201's password:
Number of key(s) added: 1
Now try logging into the machine, with:   "ssh 'server201'"
and check to make sure that only the key(s) you wanted were added.
```

此命令执行以后，会在~/.ssh 目录下多出一个用于认证的文件，其中保存了某个主机可以登录的公钥信息，这个文件为~/.ssh/authorized_keys。如果读者感兴趣，可以使用 cat 命令查看这个文件中的内容，此文件中的内容就是 id_rsa.pub 文件中的内容。

现在再使用 ssh server201 登录本机，将会发现不用输入密码即可直接登录成功。

```
[hadoop@server201 ~]$ ssh server201
Last login: Tue Mar  9 20:52:56 2021 from 192.168.56.1
```

2.2.3 Hadoop 伪分布式的安装

经过上面环境的设置，我们已经可以正式安装 Hadoop 伪分布式了。在安装之前，先确定已经安装了 JDK1.8，并正确配置了 JAVA_HOME、PATH 环境变量。接下来在磁盘根目录（/）下创建一个工作目录/app，方便我们以 hadoop 账户安装、配置与运行 Hadoop 相关程序。

步骤01 切换到根目录下：

```
[hadoop@server201 ~]# cd /
```

步骤02 添加sudo前缀，使用mkdir创建/app目录：

```
[hadoop@server201 /]# sudo mkdir /app
[sudo] hadoop 的密码:
```

步骤03 将/app目录的所有权授予hadoop用户和hadoop组：

```
[hadoop@server201 /]# sudo chown hadoop:hadoop /app
```

步骤04 su hadoop账户，切换进入/app目录：

```
[hadoop@server201 /]$ cd /app/
```

步骤05 使用ll -d命令查看本目录的详细信息：

```
[hadoop@server201 app]$ ll -d
drwxr-xr-x 2 hadoop hadoop 6 3月   9 21:35 .
```

可见此目录已经属于hadoop用户。

步骤06 将Hadoop压缩包上传到/app目录下，并解压。

步骤07 使用ll命令查看本目录：

```
[hadoop@server201 app]$ ll
总用量 386184
-rw-rw-r-- 1 hadoop hadoop 395448622 3月   9 21:40 hadoop-3.2.3.tar.gz
```

已经存在hadoop-3.2.3.tar.gz文件。

步骤08 使用tar -zxvf命令解压hadoop-3.2.3.tar.gz文件：

```
[hadoop@server201 app]$ tar -zxvf hadoop-3.2.3.tar.gz
```

步骤09 查看/app目录，已经多出hadoop-3.2.3目录：

```
[hadoop@server201 app]$ ll
总用量 386184
drwxr-xr-x 9 hadoop hadoop       149 1月   3 18:11 hadoop-3.2.3
-rw-rw-r-- 1 hadoop hadoop 395448622 3月   9 21:40 hadoop-3.2.3.tar.gz
```

步骤10 删除hadoop-3.2.3.tar.gz文件（此文件已不再需要）：

```
[hadoop@server201 app]$ rm -rf hadoop-3.2.3.tar.gz
```

步骤11 下面开始配置Hadoop。Hadoop的所有配置文件都在hadoop-3.2.3/etc/hadoop目录下，首先切换到此目录下，然后开始配置。

```
[hadoop@server201 hadoop-3.2.3]$ cd /app/hadoop-3.2.3/etc/hadoop/
```

在Hadoop的官网上有关于伪分布式配置的完整教程，它的网址是：

```
https://hadoop.apache.org/docs/stable/hadoop-project-dist/hadoop-common/Sin
gleCluster.html#Configuration
```

读者也可以根据此教程进行Hadoop伪分布式的配置学习。

1. 配置 hadoop-env.sh 文件

hadoop-env.sh 文件是 Hadoop 的环境文件，在此文件中需要配置 JAVA_HOME 变量。在此文件的第 55 行，输入以下配置，然后按 ESC 键再输入“:wq”保存退出即可。

```
export JAVA_HOME=/usr/java/jdk1.8.0_281
```

2. 配置 core-site.xml 文件

core-site.xml 文件为 HDFS 的核心配置文件，用于配置 HDFS 的协议、端口号和地址。

注意： Hadoop 3.0 以后 HDFS 的端口号建议为 8020，但如果查看 Hadoop 官网的示例，依然延续的是 Hadoop 2 之前的端口 9000，以下配置笔者将使用 8020 端口，只要保证配置的端口没有被占用即可。配置时，注意字母大小写。

使用 vim 打开 core-site.xml 文件，进入编辑模式：

```
[hadoop@server201 hadoop]$ vim core-site.xml
```

在<configuration></configuration>两个标签之间输入以下内容：

```
<property>
    <name>fs.defaultFS</name>
    <value>hdfs://server201:8020</value>
</property>
<property>
    <name>hadoop.tmp.dir</name>
    <value>/app/datas/hadoop</value>
</property>
```

配置说明：

（1）fs.defaultFS 用于配置 HDFS 的主协议，默认为 file:///。

（2）hadoop.tmp.dir 用于指定 NameNode 日志及数据的存储目录，默认为/tmp。需要使用 hadoop 账户在/app 目录下依次创建好 datas、hadoop 目录，与此配置对应。

3. 配置 hdfs-site.xml 文件

hdfs-site.xml 文件用于配置 HDFS 的存储信息。使用 vim 打开 hdfs-site.xml 文件，并在<configuration></configuration>标签中输入以下内容：

```
<property>
    <name>dfs.namenode.name.dir</name>
    <value>/app/hadoop-3.2.3/dfs/name</value>
</property>
<property>
    <name>dfs.datanode.data.dir</name>
    <value>/app/hadoop-3.2.3/dfs/data</value>
</property>
<property>
    <name>dfs.replication</name>
    <value>1</value>
</property>
<property>
```

```
        <name>dfs.permissions.enabled</name>
        <value>false</value>
    </property>
```

配置说明：

（1）dfs.replication 用于指定文件块的副本数量。HDFS 特别适合存储大文件，它会将大文件切分成每 128MB 一块，并存储到不同的 DataNode 节点上，且默认会为每一块备份 2 份，共 3 份，即此配置的默认值为 3，最大为 512。由于我们只有一个 DataNode，因此这里将文件副本数量修改为 1。

（2）dfs.permissions.enabled 用于指定访问时是否检查安全，默认为 true，为了方便访问，暂时修改为 false。

4. 配置 mapred-site.xml 文件

通过名称可见，此文件是用于配置 MapReduce 的配置文件。使用 vim 打开此文件，并在<configuration>标签中输入以下配置信息：

```
    <property>
        <name>mapreduce.framework.name</name>
        <value>yarn</value>
    </property>
```

配置说明：

mapreduce.framework.name 用于指定调试方式，这里指定使用 YARN 作为任务调用方式。

5. 配置 yarn-site.xml 文件

由于上面指定了使用 YARN 作为任务调度，因此这里需要配置 YARN 的配置信息，同样使用 vim 编辑 yarn-site.xml 文件，并在<configuration>标签中输入以下内容：

```
    <property>
        <name>yarn.resourcemanager.hostname</name>
        <value>server201</value>
    </property>
    <property>
        <name>yarn.nodemanager.aux-services</name>
        <value>mapreduce_shuffle</value>
    </property>
```

通过 hadoop classpath 命令获取所有 classpath 的目录，然后配置到上述文件中。由于还没有配置 Hadoop 的环境变量，因此需要输入完整的 Hadoop 运行路径：

```
    [hadoop@server201 hadoop]$ /app/hadoop-3.2.3/bin/hadoop classpath
```

命令执行完成后，将显示所有 classpath 信息：

```
    /home/hadoop/program/hadoop-3.2.3/etc/hadoop:/home/hadoop/program/hadoop-3.
2.3/share/hadoop/common/lib/*:/home/hadoop/program/hadoop-3.2.3/share/hadoop/co
mmon/*:/home/hadoop/program/hadoop-3.2.3/share/hadoop/hdfs:/home/hadoop/program
/hadoop-3.2.3/share/hadoop/hdfs/lib/*:/home/hadoop/program/hadoop-3.2.3/share/h
adoop/hdfs/*:/home/hadoop/program/hadoop-3.2.3/share/hadoop/mapreduce/lib/*:/ho
me/hadoop/program/hadoop-3.2.3/share/hadoop/mapreduce/*:/home/hadoop/program/ha
```

```
doop-3.2.3/share/hadoop/yarn:/home/hadoop/program/hadoop-3.2.3/share/hadoop/yar
n/lib/*:/home/hadoop/program/hadoop-3.2.3/share/hadoop/yarn/*
```

将上述信息复制一下，并使用 MobaXterm 的 XFTP 功能配置到 yarn-site.xml 文件中：

```
<property>
    <name>yarn.application.classpath</name>
    <value>/home/hadoop/program/hadoop-3.2.3/etc/hadoop:/home/hadoop/program
/hadoop-3.2.3/share/hadoop/common/lib/*:/home/hadoop/program/hadoop-3.2.3/share
/hadoop/common/*:/home/hadoop/program/hadoop-3.2.3/share/hadoop/hdfs:/home/hado
op/program/hadoop-3.2.3/share/hadoop/hdfs/lib/*:/home/hadoop/program/hadoop-3.2
.3/share/hadoop/hdfs/*:/home/hadoop/program/hadoop-3.2.3/share/hadoop/mapreduce
/lib/*:/home/hadoop/program/hadoop-3.2.3/share/hadoop/mapreduce/*:/home/hadoop/
program/hadoop-3.2.3/share/hadoop/yarn:/home/hadoop/program/hadoop-3.2.3/share/
hadoop/yarn/lib/*:/home/hadoop/program/hadoop-3.2.3/share/hadoop/yarn/*</value>
</property>
```

配置说明：

（1）yarn.resourcemanager.hostname 用于指定 ResourceManger 的运行主机，默认为 0.0.0.0，即本机。

（2）yarn.nodemanager.aux-services 用于指定执行计算的方式为 mapreduce_shuffle。

（3）yarn.application.classpath 用于指定运算时的类加载目录。

注意：这个配置的<value>标签及其值需要在同一行上，否则运行 Hadoop 会出错。

6. 配置 workers 文件

workers 文件在以前的版本中叫作 slaves，它们所起的作用一样。workers 文件主要用于在启动 Hadoop 时启动 DataNode 和 NodeManager。

编辑 workers 文件，并输入本地主机名称：

```
server201
```

7. 配置 Hadoop 环境变量

编辑/etc/profile 文件：

```
$ sudo vim /etc/profile
```

在文件里添加以下内容：

```
export HADOOP_HOME=/app/hadoop-3.2.3
export PATH=$PATH:$HADOOP_HOME/bin
```

使用 source 命令让环境变量生效：

```
$ source /etc/profile
```

使用 hdfs version 命令查看环境变量是否生效，如果配置成功，则会显示 Hadoop 的版本：

```
[hadoop@server201 hadoop]$ hdfs version
Hadoop 3.2.3
Source code repository Unknown -r 7a3bc90b05f257c8ace2f76d74264906f0f7a932
Compiled by hexiaoqiao on 2021-01-03T09:26Z
Compiled with protoc 2.5.0
```

```
    From source with checksum 5a8f564f46624254b27f6a33126ff4
    This command was run using
/app/hadoop-3.2.3/share/hadoop/common/hadoop-common-3.2.3.jar
```

8. 初始化 Hadoop 的文件系统

在使用 Hadoop 之前，必须先初始化 HDFS 文件系统，初始化的文件系统将会生成在 hadoop.tmp.dir 配置的目录下，即上面配置的/app/datas/hadoop 目录下：

```
$ hdfs namenode -format
```

命令执行完成后，若在输出的日志中找到以下这句话，即可确认初始化成功：

```
Storage directory /opt/hadoop_tmp_dir/dfs/name has been successfully formatted.
```

9. 启动 HDFS 和 YARN

启动和停止 HDFS 及 YARN 的脚本在$HADOOP_HOME/sbin 目录下，其中 start-dfs.sh 为启动 HDFS 的脚本，start-yarn.sh 为启动 ResourceManager 的脚本。以下分别启动 HDFS 和 YARN：

```
[hadoop@server201 /]$ /app/hadoop-3.2.3/sbin/start-dfs.sh
[hadoop@server201 /]$ /app/hadoop-3.2.3/sbin/start-yarn.sh
```

启动完成以后，通过 jps 命令来查看 Java 进程快照，我们会发现有 5 个进程正在运行：

```
[hadoop@server201 /]$ jps
12369 NodeManager
12247 ResourceManager
11704 NameNode
12025 SecondaryNameNode
12686 Jps
11839 DataNode
```

其中 NameNode、SecondaryNameNode、DataNode 是通过 start-dfs.sh 脚本启动的，ResourceManager 和 NodeManager 是通过 start-yarn.sh 脚本启动的。

在启动 HDFS 和 YARN 成功以后，也可以通过 http://server201:9870（宿主机上使用 http://192.168.56.201:9870）查看 NameNode 的信息，如图 2-34 所示。

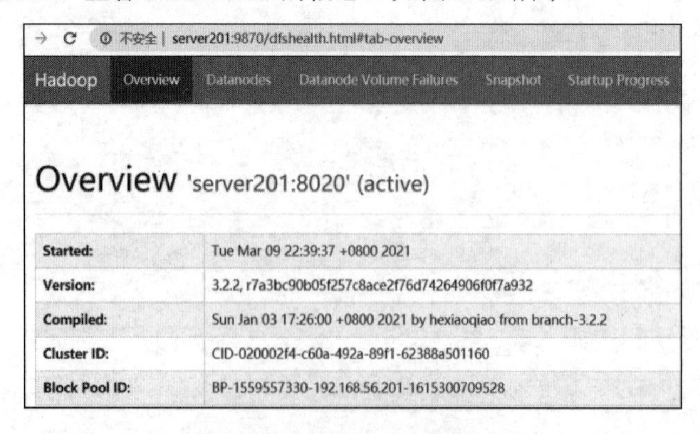

图 2-34　HDFS 的 Web 界面

可以通过 http://server201:8088（宿主机上使用 http://192.168.56.201:8088）查看 MapReduce 的信息，如图 2-35 所示。

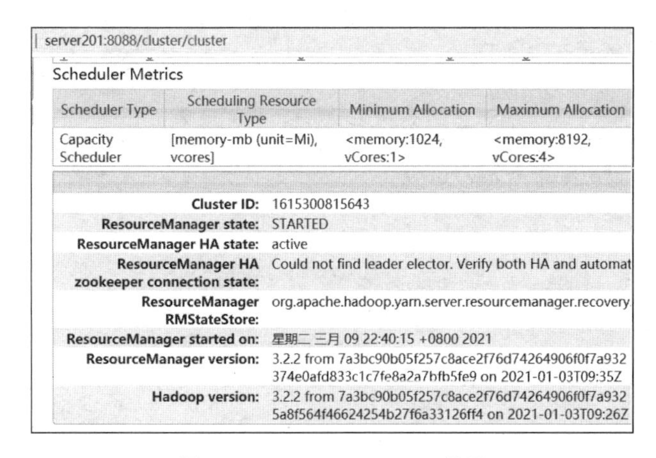

图 2-35　MapReduce Web 界面

10. 关闭 HDFS 和 YARN

执行 stop-dfs.sh 和 stop-yarn.sh 命令关闭 HDFS 和 YARN：

```
[hadoop@server201 /]$ /app/hadoop-3.2.3/sbin/stop-yarn.sh
Stopping nodemanagers
Stopping resourcemanager
[hadoop@server201 /]$ /app/hadoop-3.2.3/sbin/stop-dfs.sh
Stopping namenodes on [server201]
Stopping datanodes
Stopping secondary namenodes [server201]
```

至此，Hadoop 单机即伪分布式运行模式安装并配置成功。但是万里长征，我们这才是小小的一步。下面将继续介绍 Hadoop 完全分布式环境的搭建。

提示：本书使用 Hadoop 伪分布式环境进行 Hive 的学习，因此，可以先跳过 2.3 节内容，待后续内容全部学习完成后，再回头学习 Hadoop 完全分布式环境。

2.3　Hadoop 完全分布式环境的搭建

2.3.1　Hadoop 完全分布式集群的搭建

在 Hadoop 的集群中，有一个 NameNode，一个 ResourceManager。在高可靠的集群环境中，可以拥有两个 NameNode 和两个 ResourceManager；在 Hadoop 3 以后，同一个 NameService 可以拥有 3 个 NameNode，这将在后面的章节中讲解。由于 NameNode 和 ResourceManager 是两个主要的服务，因此建议将它们部署到不同的服务器上。

下面我们以 3 台服务器为例，来快速学习 Hadoop 的完全分布式集群的安装，这对深入了解 Hadoop 集群运行的基本原理非常有用。以下将分步骤为大家详细讲解如何搭建 Hadoop 的完全分布式集群。

注意：可以利用虚拟机软件 VirtualBox 复制出来的 CentOS 镜像文件，快速搭建 3 个 CentOS 虚拟主机来做集群。

完整的集群主机配置如表 2-1 所示。

表 2-1 集群主机配置表

IP 地址/主机名	虚 拟 机	进 程	软 件
192.168.56.101/server101	CentOS7-101 8GB 内存，2 核	NameNode SecondaryNameNode ResourceManager DataNode NodeManager	JDK HADOOP
192.168.56.102/server102	CentOS7-102 2GB+内存，1 核	DataNode NodeManager	JDK HADOOP
192.168.56.103/server103	CentOS7-103 2GB+内存，1 核	DataNode NodeManager	JDK Hadoop

从表 2-1 中可以看出，server101 运行的进程比较多，且 NameNode 运行在上面，因此这台主机需要更多的内存。

要使用 3 台 Linux 服务器来搭建集群环境，需要完成以下工作。

步骤 01 准备工作。

（1）所有主机安装 JDK1.8+。建议将 JDK 安装到不同主机的相同目录下，这样可以减少修改配置文件的次数。

（2）在主节点（即执行 start-dfs.sh 和 start-yarn.sh 的主机）上对所有其他主机做 SSH 免密码登录。

（3）修改所有主机的主机名称和 IP 地址。

（4）配置所有主机的 hosts 文件，添加主机名和 IP 地址的映射：

```
192.168.56.101 server101
192.168.56.102 server102
192.168.56.103 server103
```

（5）使用以下命令关闭所有主机上的防火墙：

```
systemctl stop firewalld
systemctl disable firewalld
```

步骤 02 在 server101 上安装 Hadoop。

可以将 Hadoop 安装到任意的目录下，如在根目录下创建/app 然后授予 hadoop 用户即可。

将 hadoop-3.2.3.tar.gz 解压到/app 目录下，并配置/app 目录属于 hadoop 用户：

```
$ sudo tar -zxvf hadoop3.2.3.tag.gz -C /app/
```

将/app 目录及子目录授权给 hadoop 用户和 hadoop 组：

```
$suto chown hadoop:hadoop -R /app
```

接下来的配置文件都在/app/hadoop-3.2.3/etc/hadoop 目录下。配置文件 hadoop-env.sh：

```
export JAVA_HOME=/usr/java/jdk1.8.0_281
```

配置文件 core-site.xml：

```
<configuration>
    <property>
        <name>fs.defaultFS</name>
        <value>hdfs://server101:8020</value>
    </property>
    <property>
        <name>hadoop.tmp.dir</name>
        <value>/app/datas/hadoop</value>
    </property>
</configuration>
```

配置文件 hdfs-site.xml：

```
<configuration>
    <property>
        <name>dfs.namenode.name.dir</name>
        <value>/app/hadoop-3.2.3/dfs/name</value>
    </property>
    <property>
        <name>dfs.datanode.data.dir</name>
        <value>/app/hadoop-3.2.3/dfs/data</value>
    </property>
    <property>
        <name>dfs.replication</name>
        <value>3</value>
    </property>
    <property>
        <name>dfs.permissions.enabled</name>
        <value>false</value>
    </property>
</configuration>
```

配置文件 mapred-site.xml：

```
<configuration>
    <property>
        <name>mapreduce.framework.name</name>
        <value>yarn</value>
    </property>
</configuration>
```

配置文件 yarn-site.xml：

```
<configuration>
    <property>
        <name>yarn.nodemanager.aux-services</name>
        <value>mapreduce_shuffle</value>
    </property>
    <property>
        <name>yarn.resourcemanager.hostname</name>
        <value>server101</value>
    </property>
    <property>
        <name>yarn.application.classpath</name>
```

```
        <value>请自行执行 hadoop classpath 命令并将结果填入</value>
    </property>
</configuration>
```

配置 workers 配置文件。workers 配置文件用于配置执行 DataNode 和 NodeManager 的节点：

```
server101
server102
server103
```

步骤 03 使用 scp 将 Hadoop 分发到其他主机。

由于 scp 会在网络上传递文件，而 hadoop/share/doc 目录下都是文档，没有必要进行复制，因此可以删除这个目录。

删除 doc 目录：

```
$ rm -rf /app/hadoop-3.2.3/share/doc
```

然后复制 server101 的文件到其他两台主机的相同目录下：

```
$scp -r /app/hadoop-3.2.3  server102:/app/
$scp -r /app/hadoop-3.2.3  server103:/app/
```

步骤 04 在 server101 上格式化 NameNode。

首先需要在 server101 上配置 Hadoop 的环境变量，打开/etc/profile 文件：

```
$ sudo vim /etc/profile
```

在文件最后追加：

```
export HADOOP_HOME=/app/hadoop-3.2.3
export PATH=$PATH:$HADOOP_HOME/bin
```

在 server101 上执行 namenode 初始化命令：

```
$ hdfs namenode -format
```

步骤 05 启动 HDFS 和 YARN。

在 server101 上执行启动工作，由于配置了集群，因此此启动过程会以 SSH 方式登录其他两台主机，并分别启动 DataNode 和 NodeManager。

```
$ /app/hadoop-3.2.3/sbin/start-dfs.sh
$ /app/hadoop-3.2.3/sbin/start-yarn.sh
```

启动完成后，通过宿主机的浏览器查看 9870 端口，页面会显示集群情况。

访问 http://192.168.56.101:9870 地址，会发现 3 个 DataNode 节点同时存在，如图 2-36 所示。

访问 http://192.168.56.101:8088 地址，会发现集群的 3 个活动节点同时存在，如图 2-37 所示。

最后，建议执行 MapReduce 测试一下集群，比如执行 WordCount 示例，如果可以顺利执行完成，则说明整个集群的配置都是正确的。命令中的 a.txt 文件内容参见 2.2.1 节。

```
$ hdfs dfs -mkdir -p /home/hadoop
$ hdfs dfs -mkdir /home/hadoop
$ hdfs dfs -put ./a.txt /home/hadoop
$ yarn jar
```

```
/app/hadoop-3.2.3/share/hadoop/mapreduce/hadoop-mapreduce-examples-3.2.3.jar
wordcount ~/a.txt /out002
```

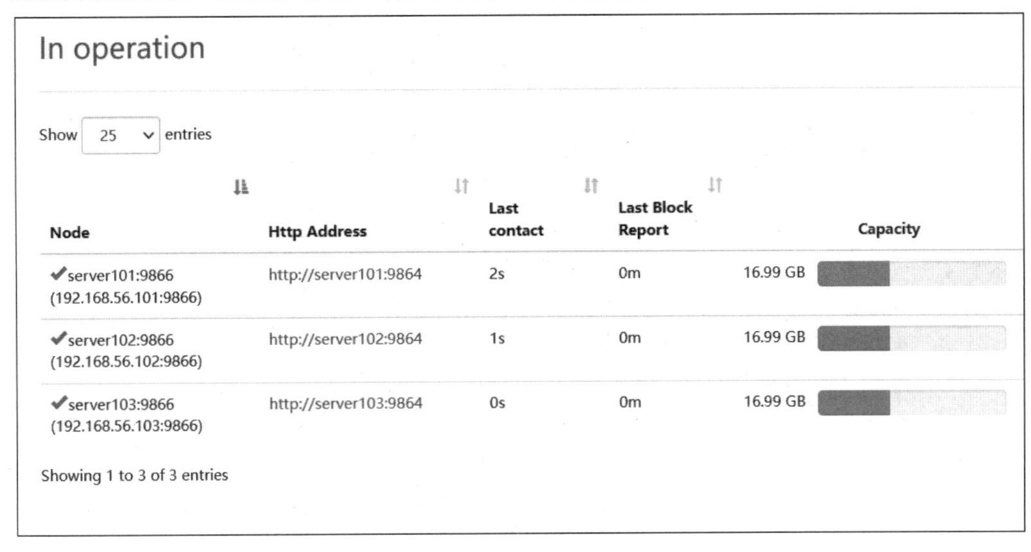

In operation

Show 25 entries

Node	Http Address	Last contact	Last Block Report	Capacity
✔server101:9866 (192.168.56.101:9866)	http://server101:9864	2s	0m	16.99 GB
✔server102:9866 (192.168.56.102:9866)	http://server102:9864	1s	0m	16.99 GB
✔server103:9866 (192.168.56.103:9866)	http://server103:9864	0s	0m	16.99 GB

Showing 1 to 3 of 3 entries

图 2-36　3 个 DataNode 节点同时存在

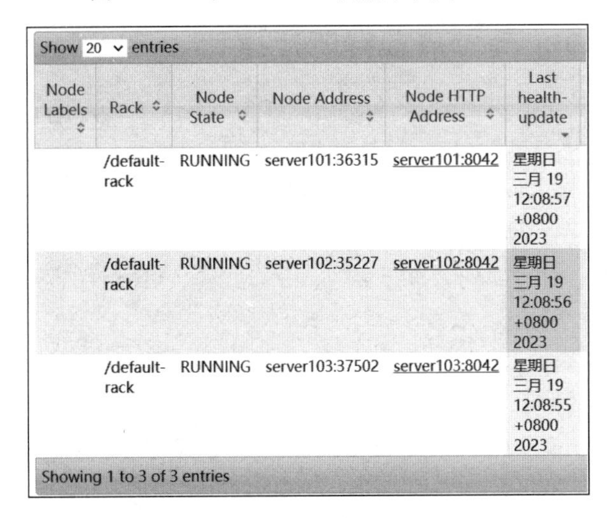

Show 20 entries

Node Labels	Rack	Node State	Node Address	Node HTTP Address	Last health-update
	/default-rack	RUNNING	server101:36315	server101:8042	星期日 三月 19 12:08:57 +0800 2023
	/default-rack	RUNNING	server102:35227	server102:8042	星期日 三月 19 12:08:56 +0800 2023
	/default-rack	RUNNING	server103:37502	server103:8042	星期日 三月 19 12:08:55 +0800 2023

Showing 1 to 3 of 3 entries

图 2-37　3 个活动节点同时存在

这里推荐读者使用 VirtualBox 把 2.2.3 节配置好的 Hadoop 伪分布式的虚拟机 CentOS7-201 复制出来，稍微做些修改，即可快速搭建 Hadoop 完全分布式环境。搭建方法如下：

（1）把 CentOS7-201 复制为 CentOS7-101，按上述步骤 01~步骤 03 核对和修改相关配置，已经配置好的可以跳过去。

（2）将 CentOS7-101 复制为 CentOS7-102、CentOS7-103，由于此时 CentOS7-101 已基本配置好了，因此复制出来的 CentOS7-102、CentOS7-103 只要改一下主机名称和 IP 地址即可。

（3）3 台虚拟机配置好了后，再按步骤 04 和步骤 05 运行这个完全分布式集群。

Hadoop 完全分布式环境如图 2-38 所示。

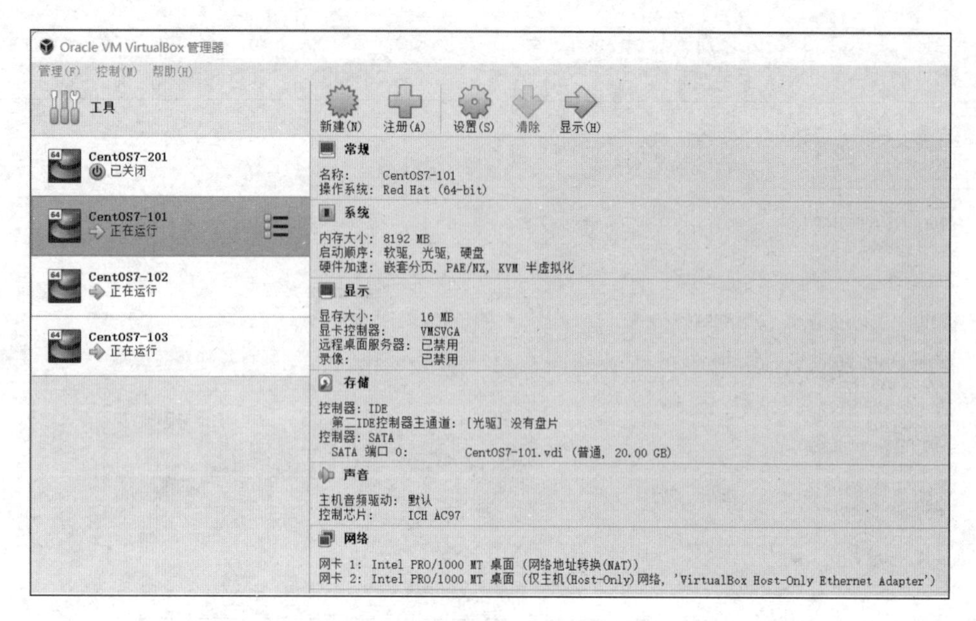

图 2-38 Hadoop 完全分布式环境

2.3.2 ZooKeeper 高可靠集群的搭建

因为高可靠集群搭建离不开 ZooKeeper，所以首先介绍一下 ZooKeeper。

ZooKeeper 是一个开放源代码的分布式应用程序协调服务，它包含一个简单的原语集，分布式应用程序可以基于它实现同步服务、配置维护和命名服务等。ZooKeeper 是 Hadoop 的一个子项目。在分布式应用中，由于工程师不能很好地使用锁机制，以及基于消息的协调机制不适合在某些应用中使用，因此需要有一种可靠的、可扩展的、分布式的、可配置的协调机制来统一系统的状态。ZooKeeper 的目的就在于此。

目前，在分布式协调技术方面做得比较好的就是 Google 的 Chubby，还有 Apache 的 ZooKeeper，它们都是分布式锁的实现者。有人会问，既然有了 Chubby 为什么还要弄一个 ZooKeeper，难道 Chubby 做得不够好吗？不是 Chubby 不够好，而是因为 Chubby 是非开源的，为 Google 自家用。后来雅虎模仿 Chubby 开发出了 ZooKeeper，也实现了类似的分布式锁的功能，并且将 ZooKeeper 作为一种开源的程序捐献给了 Apache。ZooKeeper 在分布式领域久经考验，它的可靠性、可用性都是经过理论和实践验证的，因此，在构建一些分布式系统的时候，可以将 ZooKeeper 作为起点来构建我们的系统，这将节省不少成本，而且 Bug 也会更少。

1. ZooKeeper 集群概述

（1）ZooKeeper 集群中服务器有 3 种角色：Leader、Follower 和 Observer。

- Leader：负责投票的发起与决议，更新系统状态，写数据。
- Follower：用于接收客户端请求并返回结果，在选举过程中参与投票。
- Observer：可以接收客户端连接，将写请求转发给 Leader 节点，但是不参与投票过程，只同步 Leader 状态，主要目的就是提高读取效率。

引进 Observer 角色的原因：ZooKeeper 需保证高可用和强一致性，为了支持更多的客户端，需要增加更多服务器，而服务器增多，投票阶段的延迟就会增大，影响性能；因此引入 Observer，Observer 不参与投票，只接收客户端的连接，并将写请求转发给 Leader 节点；加入更多 Observer 节点，可以提高伸缩性，同时不影响吞吐率。

建议 ZooKeeper 的集群节点个数为奇数，因为只要超过一半的机器能够正常提供服务，那么整个集群都是可用的状态，所有集群节点个数最少满足 $2n+1$（$n \geq 1$）个。

（2）ZooKeeper 的数据一致性是依靠 ZAB 协议来完成的。

ZAB（ZooKeeper Atomic Broadcast，原子广播）协议是为 ZooKeeper 特殊设计的一种支持崩溃恢复的原子广播协议。在 ZooKeeper 中，主要依赖 ZAB 协议来实现分布式数据的一致性，基于该协议，ZooKeeper 实现了一种主备模式（即 Leader 和 Follower 模型）的系统架构来保持集群中各个副本之间的数据一致性。

ZAB 协议有两种模式，分别是崩溃恢复和消息广播。当整个 ZooKeeper 集群刚刚启动或者 Leader 服务器宕机、重启或者网络故障导致没有过半的服务器与 Leader 服务器保持正常通信时，所有服务器进入崩溃恢复模式。首先选举产生新的 Leader 服务器，然后集群中的 Follower 服务器开始与新的 Leader 服务器进行数据同步，当集群中超过半数机器与该 Leader 服务器完成数据同步之后，退出恢复模式进入消息广播模式，Leader 服务器开始接收客户端的事务请求，生成事务提案来进行处理。

2. ZooKeeper 的配置文件

ZooKeeper 的默认配置文件为$ZOOKEEPER_HOME/conf/zoo.cfg，这个文件的主要配置内容如下：

```
tickTime=2000
initLimit=10
syncLimit=5
dataDir=/temp/data
clientPort=2181
#maxClientCnxns=60
#autopurge.snapRetainCount=3
#autopurge.purgeInterval=1
server.101=192.168.56.101:2888:3888
server.102=192.168.56.102:2888:3888
server.103=192.168.56.103:2888:3888
```

其中主要配置项的含义解释如下：

1）tickTime：CS 通信心跳时间

ZooKeeper 服务器之间或客户端与服务器之间维持心跳的时间间隔，也就是每过 tickTime 时间就会发送一个心跳。tickTime 以毫秒为单位，如 tickTime=2000。

2）initLimit：LF 初始通信时限

集群中的 Follower 服务器（F）与 Leader 服务器（L）之间初始连接时能容忍的最多心跳数（tickTime 的数量），如 initLimit=5。

3）syncLimit：LF 同步通信时限

集群中的 Follower 服务器与 Leader 服务器的请求和应答之间能容忍的最多心跳数（tickTime 的数量），如 syncLimit=2。

4）dataDir：数据文件目录

ZooKeeper 保存数据的目录。默认情况下，ZooKeeper 将写数据的日志文件也保存在这个目录里，如设置 dataDir=/home/zoo/SomeData。

5）clientPort：客户端连接端口

客户端连接 ZooKeeper 服务器的端口，ZooKeeper 会监听这个端口，接收客户端的访问请求，如 clientPort=2181。

6）服务器名称与地址：集群信息

集群信息包括服务器编号、服务器地址、LF 通信端口、选举端口。这个配置项的书写格式比较特殊，规则如下：

```
server.N=YYY:A:B
```

如上面 zoo.cfg 配置文件中对 server.101、server.102、server.103 的配置。

3. 配置 ZooKeeper 高可靠集群

如果要搭建 Hadoop 高可靠集群，则至少需要在 3 台服务器上搭建 ZooKeeper 集群。这里可以直接使用上一小节 Hadoop 完全分布式环境进行配置。要搭建 ZooKeeper 集群，首先需在多台机器上安装 ZooKeeper，为了便于记忆，可以将 ZooKeeper 安装到相同的目录下，如/app 目录下。

在每一个 ZooKeeper 的 dataDir 目录下，创建一个 myid 文件，里面保存的是当前 ZooKeeper 节点的 id。id 不一定从 1 开始。本示例中将主机 IP 地址的最后一段作为 id，以便于记忆。

步骤 01 安装 ZooKeeper。

准备 3 台 CentOS 7 虚拟主机，这里直接使用上一小节配置好的 Hadoop 完全分布式环境。到 Apache 官方网站下载 ZooKeeper 软件包，选取当前最新发布版本 3.8.1，下载下来的文件名是 apache-zookeeper-3.8.1-bin.tar.gz。再选取 server101 虚拟机，把 ZooKeeper 安装文件上传到/app 目录下，解压并改名：

```
[hadoop@server101 app]$ tar -zxvf apache-zookeeper-3.8.1-bin.tar.gz
[hadoop@server101 app]$ mv mv apache-zookeeper-3.8.1-bin zookeeper-3.8.1
```

步骤 02 配置环境变量$ZOOKEEPER_HOME 与 zoo.cfg。

在/etc/profile 中加入$ZOOKEEPER_HOME，并使之生效：

```
#vim /etc/profile
export ZOOKEEPER_HOME=/app/zookeeper-3.8.1
export PATH=.:$PATH:$ZOOKEEPER_HOME/bin
#source /etc/profile
```

修改$ZOOKEEPER_HOME/conf/zoo.cfg，在文件最后追加以下配置，用来配置 ZooKeeper 数据保存目录：

```
dataDir=/app/datas/zk
```

配置 ZooKeeper 集群（注意 3 台 CentOS 7 主机的 IP 地址）：

```
server.101=192.168.56.101:2888:3888
```

```
server.102=192.168.56.102:2888:3888
server.103=192.168.56.103:2888:3888
```

步骤 03 把 zookeeper-3.8.1 目录复制到其他两台虚拟机上。

使用 scp 将 zookeeper-3.8.1 目录发送到其他两台虚拟机上：

```
$ scp -r zookeeper-3.8.1  server102:/app/
$ scp -r zookeeper-3.8.1  server103:/app/
```

在其他两台虚拟机上修改/etc/profile 文件，加入$ZOOKEEPER_HOME，并使之生效。

步骤 04 修改 3 台虚拟主机 dataDir 目录下的 myid 文件。

在 server101 主机上的 myid 中添加 101（这里是为了方便记忆才取此 id 的，此 id 等于 IP 地址，建议读者也按这个规则命名），即：

```
$ echo 101 > /app/datas/zk/myid
```

以此类推，分别设置其他两台主机的 myid 为 102、103：

```
$ echo 102 > /app/datas/zk/myid
$ echo 103 > /app/datas/zk/myid
```

步骤 05 分别启动 3 台主机上的 ZooKeeper。

启动 ZooKeeper：

```
$ ./zkServer.sh start
```

使用 status 检查状态：

```
[hadoop@server101 bin]$ ./zkServer.sh status
Mode: leader   #这是 leader，其他两台为 follower
```

现在就可以同步测试了。在一台主机上进行测试操作，查看其他两台主机的同步情况。

步骤 06 同步测试操作。

在一台主机上登录客户端：

```
$ zkCli.sh
```

显示当前所有目录：

```
[hadoop: localhost 2181] ls /
[zookeeper]
```

创建一个新的目录，并写入数据：

```
[hadoop: localhost : 2181 ] create /test TestData
```

再次显示当前根目录下的所有数据。也可以登录其他主机查看，如果查看到相同的结果，则表示测试信息已经同步了。

```
[hadoop: localhost : 2181 ] ls /
[test, zookeeper]
```

ZooKeeper 的命令很多，可以使用 help 查看所有可使用的命令。

```
[zhadoopk: localhost : 2181] help
```

通过上面的配置可以看出，ZooKeeper 的集群配置相对比较简单，只要配置 zoo.cfg 并指定所有服务器节点，然后在每一个节点的 data 目录下将当前 id 写入 myid 文件中即可。

2.3.3 Hadoop 高可靠集群的搭建

搭建 Hadoop 高可靠集群，将会启动一些新的进程，它们是：

- ZKFC（DFSZKFailoverController）：Hadoop 进程，ZooKeeper 通过与 ZKFC 通信获取 NameNode 的活动状态。如果 ZKFC 认为 NameNode 已经宕机，就会通过 ZooKeeper 开启新的选举程序，选举出新的主 NameNode。
- QJM（Quorum Journal Manager）：用于同步主 NameNode 的日志数据保存到备份 NameNode 中。在高可靠情况下，将不再使用 SecondaryNameNode 程序，多个 NameNode 之间互相备份。QJM 负责同步它们的通信数据。

由于配置 Hadoop 的高可靠集群比较复杂，因此，我们需要做好配置前的规划。

- 配置从一台主机到其他主机的 SSH 免密码登录。主要是执行 start-dfs.sh 和 start-yarn.sh 的主机到其他主机的 SSH 免密码登录。
- 关闭所有主机的防火墙，仅非生产环境。
- 配置所有主机的静态地址和 hosts 文件。
- 所有主机上安装 JDK1.8+，并配置 Java 环境变量。
- 至少在 3 台主机上安装好 ZooKeeper，并启动 ZooKeeper 集群。
- 在一台主机上配置好 Hadoop 的所有配置文件，并通过 scp 分发到所有主机的相同目录下。
- 所有主机配置 Hadoop 环境变量。
- 启动 JournalNode。
- 在某台配置了 NameNode 的主机上格式化 NameNode，然后将格式化后的目录复制到其他的主机（注意只是指配置了 NameNode 的主机）上。
- 格式化 ZKFC。
- 启动 HDFS，启动 YARN。

步骤 01 准备配置计划和 3 台 CentOS 7 主机。

准备 3 台 CentOS 7 主机，配置计划如表 2-2 所示。

表 2-2 高可靠集群主机配置表

IP 地址/主机名	软 件	进 程
192.168.56.101 server101	JDK1.8+ Zookeeper3.6.2 Hadoop3.2.3	QuorumPeerMain NameNode ZKFC QJM ResourceManager NodeManager DataNode

（续表）

IP 地址/主机名	软　件	进　程
192.168.56.102 server102	JDK1.8+ Zookeeper3.6.2 Hadoop3.2.3	QuorumPeerMain NameNode ZKFC QJM ResourceManager NodeManager DataNode
192.168.56.103 server103	JDK1.8+ Zookeeper3.6.2 Hadoop3.2.3	QuourmPeerMain NameNode ZKFC QJM ResourceManager NodeManager DataNode

步骤 02 前期准备，可以继续使用上一小节的高可靠环境进行配置，所有操作都检查一下。

所有主机关闭防火墙：

```
$ sudo sytemctl stop firewalld
$ sudo systemctl disable firewalld
```

所有主机安装 JDK1.8+，并配置环境变量：

```
$ sudo tar -zxvf ~/jdk1.8-281.tag.gz -C /usr/java/
$ sudo vim /etc/profile
export JAVA_HOHE=/usr/java/jdk1.8-281
exoort PATH=$PATH:$JAVA_HOME/bin
```

所有主机设置静态 IP 地址，修改主机名称，需要分别修改 3 台主机。以下以一台主机为例：

```
$ sudo vim /etc/sysconfig/network-scripts/ifcfg-enp0s8
IPADDR=192.168.56.101
$ sudo systemctl hostnamectl set-hostname server101
$ sudo vim /etc/hosts
192.168.56.101 server101
192.168.56.102 server102
192.168.56.103 server103
```

设置所有主机 selinux=disabled：

```
$ sudo vim /etc/selinux/config
selinux=disabled
```

安装好 ZooKeeper 集群并启动。有关 ZooKeeper 集群的搭建可参看 2.3.2 节。

步骤 03 配置 hadoop-env.sh 文件。

在 hadoop-env.sh 文件中添加 JAVA_HOME 环境变量：

```
export JAVA_HOME=/usr/local/java/jdk1.8.0_281
```

步骤 04 配置 core-site.xml 文件。

配置 Hadoop 的 core-site.xml 文件与 2.2.3 节的一样，但请注意，之前配置的 fs.defaultFS 的值为 hdfs://server101:8020，在集群模式下配置的值应该为 hdfs://cluster。

```
1. <configuration>
2.      <property>
3.          <name>fs.defaultFS</name>
4.          <value>hdfs://cluster</value>
5.      </property>
6.      <property>
7.          <name>hadoop.tmp.dir</name>
8.          <value>/app/datas/hadoop</value>
9.      </property>
10.     <property>
11.         <name>ha.zookeeper.quorum</name>
12.         <value>server101:2181,server102:2181,server103:2181</value>
13.     </property>
14. </configuration>
```

配置说明：

（1）第 4 行指定集群名称，其中 cluster 可以是任意的名称，但请牢记此名称，后面将会用到。

（2）第 8 行配置 NameNode 保存数据的本地目录，此目录使用 hdfs namenode -format 格式化后将会有数据。将格式化完成的目录通过 scp 复制到其他 NameNode 节点的相同目录下，这样做是为了保证 Hadoop 集群 ID 的唯一性。

（3）第 12 行指定 ZooKeeper 的集群地址。由于我们集群的主机不是很多，因此这里的 ZooKeeper 的地址很多与 Hadoop 的存储计算结果相同。在正式的生产环境下，此 ZooKeeper 节点应该有独立的主机。

步骤 05 配置 hdfs-site.xml 文件。

这里的配置信息比较多，需注意观察。hdfs-site.xml 文件配置与 2.2.3 节中的一样，不过在集群环境下，还需要配置集群所对应的 NameService。NameService 是虚拟集群服务名称。由于配置项比较多，因此将每一个配置的具体含义都直接添加到相应配置的上面。

```
<configuration>
    <!--指定 hdfs 的 nameservice 为 cluster，需要和 core-site.xml 中的保持一致 -->
    <property>
        <name>dfs.nameservices</name>
        <value>cluster</value>
    </property>
    <!-- cluster 下面有多个 NameNode，分别取名为 nn1、nn2、nn3，也可以取其他名称。注意，
namenodes 后缀为 cluster，即之前配置的名称-->
    <property>
        <name>dfs.ha.namenodes.cluster</name>
        <value>nn1,nn2,nn3</value>
    </property>
    <!--配置每一个 NameNode 的 RPC 通信地址-->
    <property>
        <name>dfs.namenode.rpc-address.cluster.nn1</name>
        <value>server101:8020</value>
```

```
    </property>
    <property>
        <name>dfs.namenode.rpc-address.cluster.nn2</name>
        <value>server102:8020</value>
    </property>
    <property>
        <name>dfs.namenode.rpc-address.cluster.nn3</name>
        <value>server103:8020</value>
    </property>
    <!--配置每一个 NameNode 的 Web  HTTP 地址-->
    <property>
        <name>dfs.namenode.http-address.cluster.nn1</name>
        <value>server101:9870</value>
    </property>
    <property>
        <name>dfs.namenode.http-address.cluster.nn2</name>
        <value>server102:9870</value>
    </property>
    <property>
        <name>dfs.namenode.http-address.cluster.nn3</name>
        <value>server103:9870</value>
    </property>
    <!--配置 QJM 的地址-->
    <property>
        <name>dfs.namenode.shared.edits.dir</name>
        <value>
qjournal://server101:8485;server102:8485;server103:8485/cluster
        </value>
    </property>
    <!--配置 QJM 日志的目录-->
    <property>
        <name>dfs.journalnode.edits.dir</name>
        <value>/app/datas/hadoop/qjm</value>
    </property>
    <!--配置为打开自动切换功能，需要在 core-site.xml 文件中配置 ZooKeeper 地址-->
    <property>
        <name>dfs.ha.automatic-failover.enabled</name>
        <value>true</value>
    </property>
    <property>
        <name>dfs.client.failover.proxy.provider.cluster</name>
        <value>
org.apache.hadoop.hdfs.server.namenode.ha.ConfiguredFailoverProxyProvider
        </value>
    </property>
    <!--配置自动切换的方式-->
    <property>
        <name>dfs.ha.fencing.methods</name>
        <value>
                sshfence
                shell(/bin/true)
        </value>
    </property>
    <!--配置 SSH  key，注意根据不同的用户名修改目录-->
    <property>
```

```
        <name>dfs.ha.fencing.ssh.private-key-files</name>
        <value>/home/hadoop/.ssh/id_rsa</value>
    </property>
    <property>
        <name>dfs.permissions.enabled</name>
        <value>false</value>
    </property>
    <!-- 配置 sshfence 隔离机制超时时间 -->
    <property>
        <name>dfs.ha.fencing.ssh.connect-timeout</name>
        <value>30000</value>
    </property>
</configuration>
```

步骤 06 配置 mapred-site.xml。

与 2.2.3 节的配置完全一样，配置调试模式为 YARN 即可。

```
<configuration>
<!-- 指定 MapReduce 框架为 YARN 方式 -->
    <property>
        <name>mapreduce.framework.name</name>
        <value>yarn</value>
    </property>
</configuration>
```

步骤 07 配置 yarn-site.xml。

配置 yarn-site.xml 文件与配置 hdfs-site.xml 一样，需要指定集群的配置环境。为了方便阅读，我们直接将配置说明添加到每个配置功能的标签之上。

```
<configuration>
<!--开启 ResourceManager HA 功能-->
    <property>
        <name>yarn.resourcemanager.ha.enabled</name>
        <value>true</value>
    </property>
    <!-- 给 resource manager 取一个任意的名称-->
    <property>
        <name>yarn.resourcemanager.cluster-id</name>
        <value>cluster1</value>
    </property>
    <!--配置 Resourcemanager 个数，Hadoop 3 以后可以为 3 个以上，值为任意名称-->
    <property>
        <name>yarn.resourcemanager.ha.rm-ids</name>
        <value>rm1,rm2,rm3</value>
    </property>
    <!--以下配置每一个 MapReduce 的地址-->
    <property>
        <name>yarn.resourcemanager.hostname.rm1</name>
        <value>server101</value>
    </property>
    <property>
        <name>yarn.resourcemanager.hostname.rm2</name>
        <value>server102</value>
    </property>
```

```
    <property>
        <name>yarn.resourcemanager.hostname.rm3</name>
        <value>server103</value>
    </property>
    <!--配置每一个 MapReduce 的 HTTP 地址-->
    <property>
        <name>yarn.resourcemanager.webapp.address.rm1</name>
        <value>server101:8088</value>
    </property>
    <property>
        <name>yarn.resourcemanager.webapp.address.rm2</name>
        <value>server102:8088</value>
    </property>
    <property>
        <name>yarn.resourcemanager.webapp.address.rm3</name>
        <value>server103:8088</value>
    </property>
    <!--配置 ZooKeeper 地址-->
    <property>
        <name>yarn.resourcemanager.zk-address</name>
        <value>server101:2181,server102:2181,server103:2181</value>
    </property>
    <property>
        <name>yarn.nodemanager.aux-services</name>
        <value>mapreduce_shuffle</value>
    </property>
    <!—Hadoop 3 里面必须添加 classpath-->
    <property>
        <name>yarn.application.classpath</name>
        <value>请自行将 hadoop classpath 执行结果添加到这里</value>
    </property>
</configuration>
```

步骤 08 配置 workers 文件。

workers 用于指定 DataNode 节点的位置。在文件里面添加主机的名称或 IP 地址即可，一行一个：

```
server101
server102
server103
```

步骤 09 配置 Hadoop 的环境变量。

在 server101 主机上配置 Hadoop 环境变量，并通过 scp 将 profile 文件分发到其他主机上的相同目录下。

```
$ sudo vim /etc/profile
export HADOOP_HOME=/app/hadoop-3.2.3
export PATH=$PATH:$HADOOP_HOME/bin
$ sudo scp /etc/profile root@server102:/etc/
$ sudo scp /etc/profile root@server103:/etc/
```

执行以下命令让环境变量生效：

```
$source /etc/profile
```

步骤 10 复制文件到其他主机。

将配置好的 Hadoop 目录和 Hadoop 配置文件使用 scp 命令复制到其他主机的相同目录下。由于 share 目录下的 doc 里面都是帮助文件，因此可以删除这个目录，以加快复制速度。

删除 doc 目录：

```
$ rm -rf /app/hadoop-3.2.3/share/doc
```

复制文件：

```
$ scp -r /app/hadoop-3.2.3  server102:/app/
$ scp -r /app/hadoop-3.2.3  server103:/app/
```

步骤 11 启动 JournalNode。

分别在 server101、server102、server103 上执行启动 JournalNode 的命令：

```
$ /app/hadoop-3.2.3/sbin/hadoop-daemon.sh start journalnode
```

步骤 12 格式化 HDFS。

在 server101 主机上执行格式化 NameNode 的命令：

```
$ hdfs namenode -format
```

格式化后，会在 core-site.xml 中 hadoop.tmp.dir 定义的目录下生成一个文件，将这个文件使用 scp 命令复制到 server102、server103 的相同目录下。因为都是 NameNode 节点，所以必须拥有相同的数据文件。格式化成功的标志是在输出的日志中查看到以下语句：

```
Storage directory /opt/hadoop_tmp_dir/dfs/name has been successfully formatted
```

现在将格式化后的 HDFS 目录复制到 server102、server103 主机的相同目录下：

```
$ scp -r /app/datas/hadoop  server102:/app/datas/
$ scp -r /app/datas/hadoop  server103:/app/datas/
```

步骤 13 格式化 ZKFC。

在 server101 上执行格式化 ZKFC 的命令：

```
$ hdfs zkfc -formatZK
Successfully created /hadoop-ha/cluster in ZK.
```

在格式化完成以后，通过 zkCli.sh 登录 ZooKeeper 并查看目录列表，如果结果显示有一个 hadoop-ha 目录，则表示初始化成功：

```
[zk: localhost:2181(CONNECTED) 0] ls /
[zookeeper, hadoop-ha]
```

步骤 14 启动 HDFS。

在 server101 上启动 HDFS（即 NameNode）：

```
$ /app/hadoop-3.2.3/sbin/start-dfs.sh
```

此命令会根据配置同时将 server102、server103 上的 NameNode 一并启动。现在就可以通过 haadmin 命令查看整个集群的情况，此命令会输出集群中 NameNode 节点的活动状态，其中 active

为活动节点，standby 为备份节点。

```
$ hdfs haadmin -getAllServiceState
server101:8020      active
server102:8020      standby
server103:8020      standby
```

步骤 15　启动 YARN。

在 server101 上执行启动 YARN 的命令：

```
$ /app/hadoop-3.2.3/sbin/start-yarn.sh
```

集群会根据配置信息，同时启动 server102 和 server103 的 ResouceManager 进程。在启动完成以后，根据之前的配置列表分别检查每一个主机上的服务是否都已经启动。如果没有，可查看错误日志以确定原因。

启动完成以后，ResourceManager 的集群状态可以使用 rmadmin 命令查看。与 NameNode 一样，active 为活动节点，standby 为准备节点。

```
$ yarn rmadmin -getAllServiceState
server101:8033      active
server102:8033      standby
server103:8033      standby
```

步骤 16　验证高可靠集群。

通过浏览器分别访问以下地址，查看每一台 HDFS 的信息，结果如图 2-39~图 2-41 所示。

```
http://192.168.56.101:9870
http://192.168.56.102:9870
http://192.168.56.103:9870
```

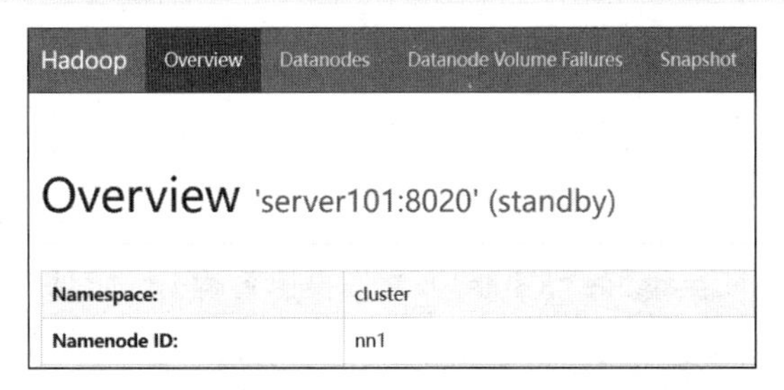

图 2-39　server101 的信息

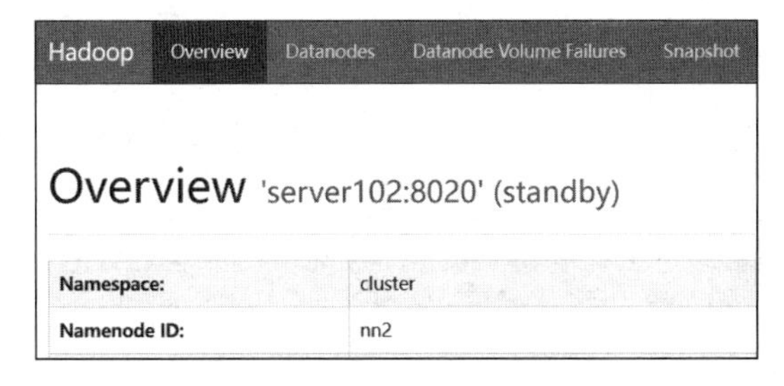

图 2-40　server102 的信息

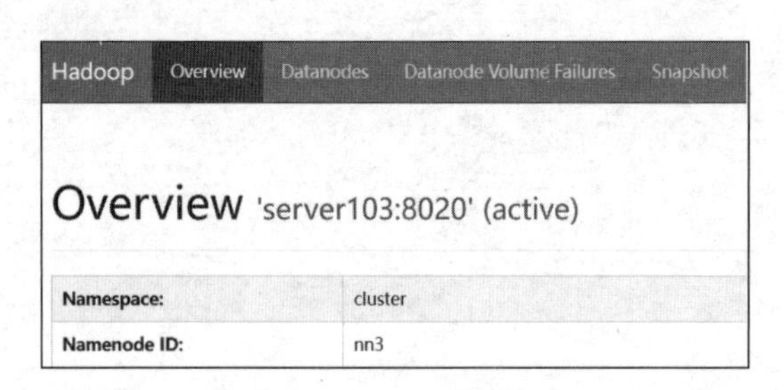

图 2-41　server103 的信息

因为 server101、server102 是准备节点，所以显示的状态为 standby，而 server103 为活动节点。也可以通过以下命令检查 NameNode 和 ResourceManager 的状态：

```
$ hdfs haadmin -getServiceState nn3
active
$ hdfs haadmin -getServiceState nn2
standby
$ yarn rmadmin -getServiceState rm3
active
$ yarn rmadmin -getServiceState rm2
standby
```

现在让我们 kill 掉 active 的 NameNode，即 kill 掉 nn3：

```
$ kill -9 <pid of nn3>
```

然后再检查状态，这时 server102 上的 NameNode 变成了 active：

```
$ hdfs haadmin -getServiceState nn2
active
```

也可以通过 haadmin 查看节点状态：

```
$ hdfs haadmin -getAllServiceState
server101   standby
server102   active
server103   failed to connected
```

手动启动那个挂掉的 NameNode，即 nn3，然后再检查状态，它已经成为 standby 了：

```
$./hadoop-daemon.sh start namenode
$ hdfs haadmin -getServiceState nn3
standby
```

使用同样的方式，可以验证 ResourceManager 是否可以自动实现容灾切换。

注意：（1）在集群配置完成以后，建议执行一个 MapReduce 测试，如 WordCount。

（2）Hadoop 的高可靠集群每一次启动相对比较麻烦，但配置成功以后，下次启动就相对比较简单了。对于上面的示例而言，再次启动只需在 server101 主机上执行./start-dfs.sh 和./start-yarn.sh 即可。

2.4　Hive 的安装与配置

由于 Hive 是运行在 Hadoop 下的数据仓库，因此必须在已经安装好 Hadoop 的环境下运行 Hive，且要正确配置 HADOOP_HOME 环境变量。本书所讲解的 Hive 运行在 2.2.2 节搭建的 Hadoop 伪分布式环境的基础上，因此我们可以利用 CentOS7-201 虚拟机（见图 2-38）进行 Hive 的安装与配置。

2.4.1　Hive 的安装与启动

步骤 01 下载、安装和配置 Hive。

Hive 官网：

```
http://hive.apache.org/
```

安装包具体下载地址：

```
http://mirror.bit.edu.cn/apache/hive/
```

这里选择下载 Hive 3.1.3 版本，文件名为 apache-hive-3.1.3-bin.tar.gz。

步骤 02 上传并解压 Hive。

把下载下来的 Hive 安装包文件上传到当前用户 hadoop 的主目录，再解压 Hive 安装包文件：

```
[hadoop@server201 app]$ tar -zxvf ~/apache-hive-3.1.3-bin.tar.gz -C .
```

目录名称太长了，修改一下名称：

```
[hadoop@server201 app]$ mv apache-hive-3.1.3-bin/ hive-3.1.3
```

配置 Hive 的环境变量是可选的，是为了方便执行 Hive 脚本：

```
export HIVE_HOME=/app/hive-3.1.3
export PATH=$PATH:$HIVE_HOME/bin
```

接下来配置 hive-site.xml，Hive 的 conf 目录下没有这个文件，我们可以自己创建一个，内容如下：

```
<?xml version="1.0" encoding="UTF-8" standalone="no"?>
<?xml-stylesheet type="text/xsl" href="configuration.xsl"?>
<configuration>
    <property>
        <name>hive.exec.scratchdir</name>
        <value>hdfs://server201:8020/hive/scratchdir</value>
    </property>
    <property>
        <name>hive.exec.local.scratchdir</name>
        <value>/app/hive-3.1.3/datas</value>
    </property>
    <property>
        <name>hive.downloaded.resources.dir</name>
        <value>/app/hive-3.1.3/datas</value></property>
    <property>
```

```
        <name>hive.scratch.dir.permission</name>
        <value>700</value>
    </property>
    <property>
        <name>hive.querylog.location</name>
        <value>/app/hive-3.1.3/datas</value>
    </property>
    <property>
        <name>hive.metastore.warehouse.dir</name>
        <value>/app/hive-3.1.3/datas</value>
    </property>
    <property>
        <name>hive.metastore.local</name>
        <value>true</value>
    </property>
    <property>
        <name>datanucleus.schema.autoCreateAll</name>
        <value>true</value>
    </property>
</configuration>
```

步骤03 启动Hadoop，登录Hive命令行。

启动Hadoop：

```
[hadoop@server201 app]$ /app/hadoop-3.2.3/sbin/start-dfs.sh
[hadoop@server201 app]$ /app/hadoop-3.2.3/sbin/start-yarn.sh
```

使用Hive脚本登录Hive命令行界面时，Hive要访问Hadoop的core-site.xml文件，并访问fs.defaultFS所指的服务器。

接下来初始化Derby元数据库，注意这个Derby是Hive内嵌的，不支持多个Hive客户端连接：

```
[hadoop@server201 app]# cd /app/hive-3.1.3/bin
[hadoop@server201 bin]#./schematool -initSchema -dbType derby
```

登录Hive的命令行：

```
[hadoop@server201 ~]$ hive
hive>
```

登录Hive的命令行后，就可以学习基本的SQL操作命令了。

2.4.2 基本的SQL操作命令

类似于MySQL的SQL命令都可以在Hive下运行。

（1）查看所有数据库：

```
hive> show databases;
OK
default
Time taken: 0.025 seconds, Fetched: 1 row(s)
```

（2）查看默认数据库下的所有表：

```
hive> show tables;
```

```
OK
Time taken: 0.035 seconds
```

（3）创建一张表，并显示这张表的结构：

```
hive> create table stud(id int,name varchar(30));
OK
Time taken: 0.175 seconds
hive> desc stud;
OK
id                  int
name                varchar(30)
Time taken: 0.193 seconds, Fetched: 2 row(s)
```

（4）显示这个张表在 Hive 中的结构：

```
hive> show create table stud;
OK
CREATE TABLE 'stud'(
  'id' int,
  'name' varchar(30))
ROW FORMAT SERDE
  'org.apache.hadoop.hive.serde2.lazy.LazySimpleSerDe'
STORED AS INPUTFORMAT
  'org.apache.hadoop.mapred.TextInputFormat'  数据存储类型
OUTPUTFORMAT
  'org.apache.hadoop.hive.ql.io.HiveIgnoreKeyTextOutputFormat' 输出类型
LOCATION
  'hdfs://server201:8020/user/hive/warehouse/stud' 保存的位置
TBLPROPERTIES (  表的其他属性信息
  'transient_lastDdlTime'='1530518761')
Time taken: 0.128 seconds, Fetched: 13 row(s)
```

（5）向表中写入一行记录。由于 Hive 将操作转成 MapReduce 程序，因此 insert 语句会被转换成 MapReduce 程序。这个效率比较低，尽量不要使用 insert 语句写入数据，而是采用 Hive 分析现有的数据。

```
hive> insert into stud values(1,'Jack');
```

运行结果中有如下内容：

```
Stage-1 map =0%,  reduce = 0%
Stage-1 map =100%,  reduce = 0%,  Cumulative CPU 2.4 sec
```

可见，一个简单的 insert 语句却执行了 MapReduce 程序，所以效率不会太高。

（6）不支持 update 和 delete：

```
hive> update stud set name='Alex' where id=1;
FAILED: SemanticException [Error 10294]: Attempt to do update or delete using transaction manager that does not support these operations.
hive> delete from stud where id=1;
FAILED: SemanticException [Error 10294]: Attempt to do update or delete using transaction manager that does not support these operations.
```

由以上运行结果可知，Hive 分析的数据是存储在 HDFS 上的，而 HDFS 不支持随机写，只支持追加写，因此在Hive中不能update和delete，只能select和insert。Hive就是用于分析现有数据的，虽然insert可以操作，但不建议使用insert语句。

2.5 Hive的一些命令

2.5.1 显示Hive的帮助

可以使用如下命令查看Hive的帮助信息：

```
[hadoop@server201 ~]$ hive --help
```

默认情况下，只要输入hive就会执行hive cli命令，即进入Hive的cli命令行模式。

hive --service命令用于可选的登录或是启动Hive的哪一种模式，默认模式为cli，即：

```
[hadoop@server201 ~]$ hive --service cli
hive>
```

2.5.2 显示Hive某个命令的帮助

以下命令显示--service cli的帮助信息：

```
[hadoop@server201 ~]$ hive --help --service cli
Hive Session ID = …
usage : hive
 -d,--define <key=value>          Variable subsitution to apply to hive
                                  commands. e.g. -d A=B or --define A=B
    --database <databasename>     Specify the database to use
 -e <quoted-query-string>         SQL from command line
 -f <filename>                    SQL from files
 -H,--help                        Print help information
    --hiveconf <property=value>   Use value for given property
    --hivevar <key=value>         Variable subsitution to apply to hive
                                  commands. e.g. --hivevar A=B
 -i <filename>                    Initialization SQL file
 -S,--silent                      Silent mode in interactive shell
 -V,--verbose                     Verbose mode (echo executed SQL to the
                                  console)
```

2.5.3 变量与属性

通过set命令可以显示Hive的所有环境变量信息，如显示当前用户的工作目录：

```
hive> set env:HOME;
env:HOME=/home/hadoop
```

又如显示当前用户的名称：

```
hive> set system:user.name;
system:user.name=hadoop
```

如果直接使用set，将显示所有环境变量信息。

以下两个命令会生成对我们有用的信息：

```
hive> set env:HADOOP_HOME
    > ;
env:HADOOP_HOME=/app/hadoop-3.2.3
hive> set env:HADOOP_CONF_DIR
    > ;
env:HADOOP_CONF_DIR=/app/hadoop-3.2.3/etc/hadoop
```

使用 set-v，将显示更多的信息。

2.5.4　指定 SQL 语句或文件

hive cli 包含-e 参数，用于指定 SQL 语句，并在最后将结果显示到控制台：

```
[hadoop@server201 ~]$ hive -e "select * from stud"
T001    Jack    1   23
T002    Mary    0   24
T003    Rose    1   34
T004    Alex    1   23
T100    Smith   1   29
T005    Jim     0   34
T000    张三    1   38
S200    Jack    1   90
S201    Tom     0   34
S301    Mary    0   NULL
S500    Rose    1   34
```

也可以将结果保存到一个文件中：

```
[hadoop@server201 ~]$ hive -e "select * from stud" >~/a.sql
```

查看结果：

```
[hadoop@server201 ~]$ cat ~/a.sql
T001    Jack    1   23
T002    Mary    0   24
T003    Rose    1   34
T004    Alex    1   23
T100    Smith   1   29
T005    Jim     0   34
T000    张三    1   38
S200    Jack    1   90
S201    Tom     0   34
S301    Mary    0   NULL
S500    Rose    1   34
```

使用过滤功能：

```
[hadoop@server201 ~]$ hive -e "set" | grep header
Logging initialized using configuration in
file:/app/hive-3.1.3/conf/hive-log4j.properties
hive.cli.print.header=false
hive.exec.rcfile.use.explicit.header=true
```

还可以通过-f 选项指定一个外部的文件，如 hive -f a.sql：

```
    Logging initialized using configuration in file:/app/hive-3.1.3/conf/hive-log4j.properties
    OK
    T001    Jack    1   23
    T002    Mary    0   24
    T003    Rose    1   34
    T004    Alex    1   23
    T100    Smith   1   29
    T005    Jim     0   34
    T000    张三    1   38
    S200    Jack    1   90
    S201    Tom     0   34
    S301    Mary    0   NULL
    S500    Rose    1   34
```

在 Hive 的命令行下，也可以使用 source 命令执行一个外部的 SQL 文件：

```
[hadoop@server201 ~]$ source ~/a.sql
```

2.5.5 显示表头

使用 set hive.cli.print.header=true 命令就可以显示表头：

```
hive> set hive.cli.print.header=true
hive> select * from stud;
OK
stud.id  stud.name stud.set  stud.age
T001    Jack    1   23
T002    Mary    0   24
T003    Rose    1   34
T004    Alex    1   23
T100    Smith   1   29
T005    Jim     0   34
T000    张三    1   38
S200    Jack    1   90
S201    Tom     0   34
S301    Mary    0   NULL
S500    Rose    1   34
```

2.6 Hive 元数据库

Hive 中的元数据包括表的名字、表的列和分区及其属性、表的属性（是否为外部表等）、表的数据所在目录等。元数据的存储主要有两种常用的方式：第一种是使用 Derby 数据库进行元数据的存储，第二种是使用 MySQL 数据库来进行 Hive 元数据的存储。

2.6.1 Derby

Hive 默认使用自带（内嵌）的 Derby 进行元数据存储，这就意味着无法实现多个 hive shell 并发连接 Hive。本小节讲解如何使用轻量级 Derby 的 C/S 服务模式来解决这个问题。

1. 下载、安装、运行 Derby 数据库

首先，从 Apache 网站下载支持 Java 8 的 Derby 版本 10.14.2.0。

```
https://db.apache.org/derby
```

下载下来的文件名是 db-derby-10.14.2.0-bin.tar.gz，使用 hadoop 账户将它上传到/app 目录，解压并改个简短的目录名：

```
tar -xzvf ./db-derby-10.14.2.0-bin.tar.gz
mv db-derby-10.14.2.0-bin/ derby-10.14.2
```

Derby 是开箱即用的，因此直接进入 Derby 的 bin 目录，启动 Derby：

```
cd derby-10.14.2/bin
./startNetworkServer -h 0.0.0.0
```

同时，还需要复制两个 JAR 包到/app/hive-3.1.3/lib 目录下：

```
cp derbyclient.jar /app/hive-3.1.3/lib/
cp derbytools.jar /app/hive-3.1.3/lib/
```

在 Hive 的配置文件 hive-site.xml 中加入如下配置信息（注意，Derby 默认可以没有用户名和密码）：

```
<property>
    <name>javax.jdo.option.ConnectionURL</name>
    <value>jdbc:derby://server201:1527/hive_meta;create=true</value>
    <description>JDBC connect string for a JDBC metastore</description>
</property>
<property>
    <name>javax.jdo.option.ConnectionDriverName</name>
    <value>org.apache.derby.jdbc.ClientDriver</value>
    <description>Driver class name for a JDBC metastore</description>
</property>
```

启动 Hadoop：

```
[hadoop@server201 app]$ /app/hadoop-3.2.3/sbin/start-dfs.sh
[hadoop@server201 app]$ /app/hadoop-3.2.3/sbin/start-yarn.sh
```

接下来初始化 Derby 元数据库：

```
[hadoop@server201 app]# cd /app/derby-10.14.2/bin
[hadoop@server201 bin]#./schematool -initSchema -dbType derby
```

2. 连接 Derby 数据库进行测试

进入 Derby 安装目录：

```
#cd /app/derby-10.14.2/bin
```

输入./ij（ij 是 Derby 的客户端工具）：

```
#./ij
ij 版本 10.4
ij> Connect'jdbc:derby://server201:1527/metastore_db;create=true'
ij> select * from INFORMATION_SCHEMA.TABLES where TABLE_TYPE = 'BASE TABLE'
```

3. Derby 元数据库的数据字典中主要的表

Derby 元数据库的数据字典中主要的表及其说明如表 2-3 所示。

表2-3 Derby元数据库的数据字典

表名	说明	关联键
COLUMNS	Hive 表字段信息（字段注释、字段名、字段类型、字段序号）	SD_ID
DBS	元数据库信息，存放 HDFS 路径信息	DB_ID
PARTITION_KEYS	Hive 分区表的分区键	PART_ID
SDS	所有 Hive 表、表分区所对应的 HDFS 数据目录和数据格式	SD_ID，SERDE_ID
SD_PARAMS	序列化反序列化信息，如行分隔符、列分隔符、NULL 的表示字符等	SERDE_ID
SEQUENCE_TABLE	SEQUENCE_TABLE 表保存了 Hive 对象的下一个可用 ID，如 'org.apache.hadoop.hive.metastore.model.MTable', 21，则下一个新创建的 Hive 表其 TBL_ID 就是 21	
TABLE_PARAMS	表级属性，如是否为外部表、表注释等	TBL_ID
TBLS	所有 Hive 表的基本信息	TBL_ID，SD_ID

表中的内容对 Hive 整个表的创建过程已经有所体现。

（1）对用户提交的 Hive 语句进行解析，将它分解为表、字段、分区等 Hive 对象。

（2）根据解析到的信息构建对应的表、字段、分区等对象，从 SEQUENCE_TABLE 中获取构建对象的最新 ID，与构建对象信息（名称、类型等）一同通过 DAO 方法写入元数据表中，成功后将 SEQUENCE_TABLE 中对应的最新 ID 加 5（+5）。

实际上我们常见的 RDBMS 都是通过这种方法进行组织的，典型的如 PostgreSQL，其系统表和 Hive 元数据一样暴露了这些 ID 信息（oid、cid 等），而 Oracle 等商业化的系统则隐藏了这些具体的 ID 信息。通过这些元数据我们可以很容易地读取数据，比如创建一个表的数据字典信息（导出建表语句）。

2.6.2 MySQL

2.4 节安装的 Hive 的元数据默认存储在内嵌的 Derby 数据库中。为了支持多用户模式，可以使用读者最常用的 MySQL 数据库存储元数据来支持并发调用 Hive，因此还可以安装 MySQL，并配置元数据到 MySQL 中。当然，读者也可以安装别的关系数据库来替代 MySQL 数据库，Hive 支持的常用数据库及其版本如表 2-4 所示。

表2-4 Hive支持的常用数据库及其版本

数据库	最小支持的版本	参数值
MySQL	5.6.17	mysql
PostgreSQL	9.1.13	postgres
Oracle	11g	oracle
MS SQL Server	2008 R2	mssql

2.7 MySQL 的安装

Hive 元数据默认存储在自带的 Derby 数据库中，不能并发调用 Hive。而 MySQL 数据库存储元数据支持多用户模式，可以并发调用 Hive，因此本节先来介绍 MySQL 的安装。

1. MySQL 安装包下载、解压及客户端安装

步骤 01 到 MySQL 官方网站下载安装包到本地，操作系统及其版本的选择如图 2-42 所示，安装包文件如图 2-43 所示。这个安装包名叫 mysql-8.0.32-1.el7.x86_64.rpm-bundle.tar，捆绑了 MySQL 客户端、MySQL 库和 MySQL 服务器文件。

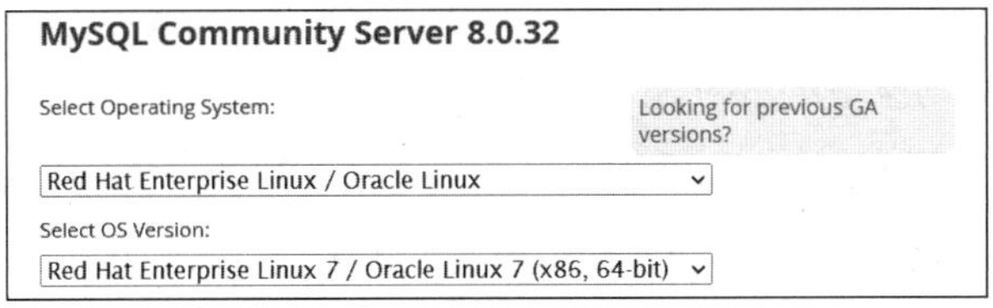

MySQL Community Server 8.0.32

Select Operating System: Looking for previous GA versions?

Red Hat Enterprise Linux / Oracle Linux

Select OS Version:

Red Hat Enterprise Linux 7 / Oracle Linux 7 (x86, 64-bit)

图 2-42 下载 MySQL

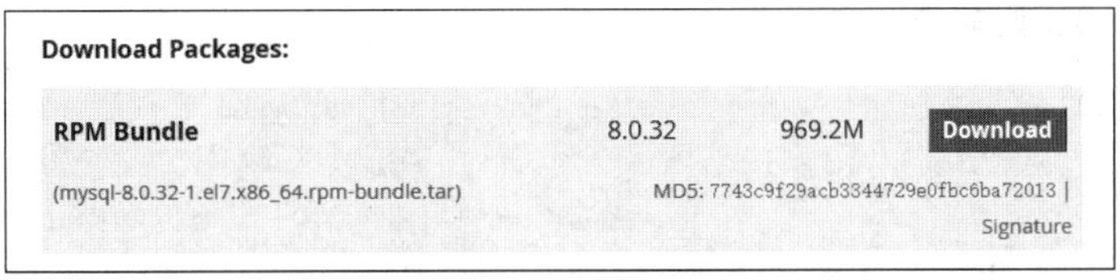

Download Packages:

RPM Bundle 8.0.32 969.2M Download

(mysql-8.0.32-1.el7.x86_64.rpm-bundle.tar) MD5: 7743c9f29acb3344729e0fbc6ba72013 | Signature

图 2-43 安装包文件

步骤 02 查看系统中 MySQL 是否已经安装了旧版本，如果安装了旧版本 MySQL 或依赖库文件，则需要卸载：

```
root@server201 ~# rpm -qa|grep -i mysql
mysql-libs-5.1.73-7.el6.x86_64
root@server201 ~# yum remove mysql-lib
```

步骤 03 把 mysql-8.0.32-1.el7.x86_64.rpm-bundle.tar 通过 MobaXterm 上传到/opt/software 目录，解压安装包文件到 software 目录下：

```
    root@server201:/opt/software# tar -xvf
mysql-8.0.32-1.el7.x86_64.rpm-bundle.tar
    mysql-8.0.32-1.el7.x86_64.rpm-bundle.tar
    mysql-community-client-8.0.32-1.el7.x86_64.rpm
    mysql-community-client-plugins-8.0.32-1.el7.x86_64.rpm
    mysql-community-common-8.0.32-1.el7.x86_64.rpm
    mysql-community-debuginfo-8.0.32-1.el7.x86_64.rpm
    mysql-community-devel-8.0.32-1.el7.x86_64.rpm
    mysql-community-embedded-compat-8.0.32-1.el7.x86_64.rpm
    mysql-community-icu-data-files-8.0.32-1.el7.x86_64.rpm
    mysql-community-libs-8.0.32-1.el7.x86_64.rpm
    mysql-community-libs-compat-8.0.32-1.el7.x86_64.rpm
    mysql-community-server-8.0.32-1.el7.x86_64.rpm
```

```
mysql-community-server-debug-8.0.32-1.el7.x86_64.rpm
mysql-community-test-8.0.32-1.el7.x86_64.rpm
```

步骤 04 分别安装通用文件、库文件等依赖包。

步骤 05 安装 MySQL 客户端程序：

```
root@server201:/opt/software/# rpm -ivh
mysql-community-client-8.0.32-1.el7.x86_64.rpm
```

2. MySQL 服务器的安装、启动与登录

步骤 01 安装 MySQL 服务器：

```
root@server201:/opt/software# rpm -ivh
mysql-community-server-8.0.32-1.el7.x86_64.rpm
```

步骤 02 查看 MySQL 自动生成的 root 临时密码：

```
root@server201:/opt/software# grep 'temporary password' /var/log/mysqld.log
```

可以查到 MySQL 管理员 root 的临时密码，记下这个密码以方便修改。比如临时密码是：

```
OEXaQuS8IWkG19Xs
```

步骤 03 启动 MySQL 服务器：

```
root@server201:/opt/software# systemctl start mysqld.service
```

步骤 04 查看 MySQL 服务器的状态：

```
root@server201:/opt/software# systemctl status mysqld.service
```

步骤 05 连接登录 MySQL：

```
root@server201:/opt/software/mysql-libs# mysql -uroot
-pOEXaQuS8IWkG19Xs
```

步骤 06 修改 root 密码，并对其他主机连接本机 MySQL 进行授权：

```
mysql>alter user 'root'@'localhost' identified by 'A1b2c3==';
mysql>grant all privileges on *.* to 'root'@'%' with grant option;
mysql>flush privileges;
```

步骤 07 退出 MySQL：

```
mysql>exit;
```

2.8 配置 MySQL 保存 Hive 元数据

如果需要支持多用户登录 Hive，则必须配置一个独立的数据库。上一节已经将 MySQL 数据库安装到 Linux 上，本节将讲解一下配置 MySQL 保存 Hive 元数据的步骤。

步骤 01 修改 hive-site.xml 文件。

继续以 MySQL 为例，首先需要在文件 hive-site.xml 中配置 MySQL 数据库的相关属性，如

表 2-5 所示。

表2-5　MySQL数据库的相关属性配置

配置参数	配置值	说明
javax.jdo.option.ConnectionURL	jdbc:mysql://<host name>/<database name>?createDatabaseIfNotExist=true	元数据存储在 MySQL 数据库上
javax.jdo.option.ConnectionDriverName	com.mysql.jdbc.Driver	MySQL JDBC 驱动
javax.jdo.option.ConnectionUserName	<user name>	连接 MySQL 数据库的用户名
javax.jdo.option.ConnectionPassword	<password>	连接 MySQL 数据库的密码

在 hive/conf 目录下有一个 hive-default.xml.template 文件，里面都是默认的配置。打开这个文件可以发现一些说明文字，如图 2-44 所示。

```
1 <?xml version="1.0" encoding="UTF-8" standalone="no"?>
2 <?xml-stylesheet type="text/xsl" href="configuration.xsl"?>
3 <configuration>
4   <!-- WARNING!!! This file is auto generated for documentation purposes ONLY! -->
5   <!-- WARNING!!! Any changes you make to this file will be ignored by Hive.   -->
6   <!-- WARNING!!! You must make your changes in hive-site.xml instead.         -->
7   <!-- Hive Execution Parameters -->
```

图 2-44　文件里有一些说明文字

意思是如果要使配置生效，就需要添加 hive-site.xml 文件，可以直接把这个文件复制为 hive-site.xml 再做配置。当然，为了简单起见，也可以直接使用 2.4.1 节中我们自己创建的 hive-site.xml。

打开 hive-site.xml 文件，将里面的内容全部删除，再添加以下配置：

```
<?xml version="1.0" encoding="UTF-8" standalone="no"?>
<?xml-stylesheet type="text/xsl" href="configuration.xsl"?>
<configuration>
   <property>
      <name>javax.jdo.option.ConnectionURL</name>
      <value>jdbc:mysql://server201:3306/hive?useSSL=false</value>
   </property>
   <property>
      <name>javax.jdo.option.ConnectionDriverName</name>
      <value>com.mysql.jdbc.Driver</value>
   </property>
   <property>
      <name>javax.jdo.option.ConnectionUserName</name>
      <value>root</value>
   </property>
   <property>
      <name>javax.jdo.option.ConnectionPassword</name>
      <value>A1b2c3==</value>
   </property>
</configuration>
```

这个配置需要我们在 MySQL 中手动创建 Hive 数据库。以下配置相同，只是多了一个 createDatabaseIfNotExist 设置，设置为 true 时表示不需要在 MySQL 中手动创建 Hive 数据库：

```
<?xml version="1.0" encoding="UTF-8" standalone="no"?>
<?xml-stylesheet type="text/xsl" href="configuration.xsl"?>
<configuration>
   <property>
      <name>javax.jdo.option.ConnectionDriverName</name>
      <value>com.mysql.jdbc.Driver</value>
   </property>
   <property>
      <name>javax.jdo.option.ConnectionURL</name>
      <value>jdbc:mysql://server201:3306/hive?createDatabaseIfNotExist
=true&useSSL=false</value>
   </property>
   <property>
      <name>javax.jdo.option.ConnectionUserName</name>
      <value>root</value>
   </property>
   <property>
      <name>javax.jdo.option.ConnectionPassword</name>
      <value>A1b2c3==</value>
   </property>
   <property>
      <name>hive.metastore.warehouse.dir</name>
      <value>/app/hive-3.1.3/warehouse</value>
   </property>
</configuration>
```

步骤 02 将 MySQL 的驱动放到/app/hive-3.1.3/lib 目录下：

```
[root@server201 lib]$ ll | grep mysql
-rw-r--r--. 1 hadoop hadoop   2480823 12月 26 2022 mysql-connector-j-8.0.32
```

步骤 03 登录 hive cli。

可以直接使用以下命令来初始化数据库：

```
[root@server80 bin]#./schematool -initSchema -dbType mysql
```

运行结果如图 2-45 所示，表示成功初始化数据库。

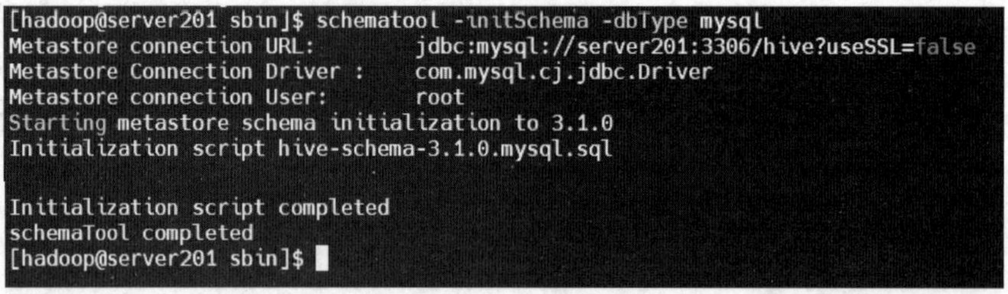

```
[hadoop@server201 sbin]$ schematool -initSchema -dbType mysql
Metastore connection URL:        jdbc:mysql://server201:3306/hive?useSSL=false
Metastore Connection Driver :    com.mysql.cj.jdbc.Driver
Metastore connection User:       root
Starting metastore schema initialization to 3.1.0
Initialization script hive-schema-3.1.0.mysql.sql


Initialization script completed
schemaTool completed
[hadoop@server201 sbin]$
```

图 2-45 运行结果

如果使用 MySQL 客户端工具登录 MySQL，就会在 MySQL 数据库的 Hive 中发现如图 2-46 所示的数据表（部分截图）。

```
mysql> show tables;
+------------------------------+
| Tables_in_hive               |
+------------------------------+
| AUX_TABLE                    |
| BUCKETING_COLS               |
| CDS                          |
| COLUMNS_V2                   |
| COMPACTION_QUEUE             |
| COMPLETED_COMPACTIONS        |
| COMPLETED_TXN_COMPONENTS     |
| CTLGS                        |
| DATABASE_PARAMS              |
| DBS                          |
| DB_PRIVS                     |
| DELEGATION_TOKENS            |
| FUNCS                        |
| FUNC_RU                      |
| GLOBAL_PRIVS                 |
| HIVE_LOCKS                   |
| IDXS                         |
| INDEX_PARAMS                 |
```

图 2-46 数据表（部分）

现在 Hive 就能支持多个 hive cli 客户端登录了。

2.9 HiveServer2 与 Beeline 配置

Hive 是大数据技术中数据仓库应用的基础组件，是其他类似数据仓库应用的对比基准。基础的数据操作可以通过脚本方式由 Hive 客户端进行处理。

若要开发应用程序，则需要使用 Hive 的 JDBC 驱动进行连接。Hive 内置了 HiveServer 和 HiveServer2 服务器，两者都允许客户端使用多种编程语言进行连接，但是 HiveServer 不能处理多个客户端的并发请求，因此产生了 HiveServer2。

HiveServer2 允许远程客户端使用各种编程语言向 Hive 提交请求并检索结果，支持多客户端并发访问和身份验证。

HiveServer2 拥有自己的 CLI，即 Beeline。Beeline 是一个基于 SQLLine 的 JDBC 客户端。由于 HiveServer2 是 Hive 官方开发维护的重点，因此推荐使用 Beeline。

步骤 01 修改 Hadoop 配置文件 core-site.xml，增加如下配置：

```
<property>
   <name>hadoop.proxyuser.root.hosts</name>
   <value>*</value>
</property>
<property>
   <name>hadoop.proxyuser.root.groups</name>
   <value>*</value>
</property>
<property>
   <name>hadoop.proxyuser.root.user</name>
   <value>*</value>
</property>
```

步骤 02 切换到/app/hive-3.1.3/conf 目录，修改 hive-site.xml 文件，写入以下配置信息：

```
<property>
```

```
   <name>hive.server2.thrift.port</name>
   <value>10000</value>
</property>
<property>
   <name>hive.server2.thrift.bind.host</name>
   <value>localhost</value>
</property>
<property>
   <name>hive.server2.thrift.port</name>
   <value>10000</value>
</property>
<property>
   <name>hive.server2.thrift.bind.host</name>
   <value>server201</value>
</property>
<property>
    <name>hive.server2.thrift.http.port</name>
    <value>10001</value>
</property>
<property>
   <name>hive.server2.enable.doAs</name>
   <value>false</value>
</property>
<!–下面配置 thrift 服务的验证账户，用于登录 HiveServer2 服务器-->
<property>
   <name>hive.server2.thrift.client.user</name>
   <value>hduser</value>
</property>
<property>
   <name>hive.server2.thrift.client.password</name>
   <value>hduser</value>
</property>
```

该配置信息配置了 HiveServer2 的端口号和主机名。

步骤 03 经过以上配置后，需要重新启动 Hadoop 集群。

步骤 04 启动 HiveServer2。以下两个命令都可以启动 HiveServer2 服务器：

```
hadoop@server201 hadoop$ hive --service hiveserver2
hadoop@server201 hadoop$ hiveserver2
```

步骤 05 使用 Beeline 连接 HiveServer2，输入命令如下：

```
hadoop@server201 hadoop$ /beeline
Beeline version 3.1.3 by Apache Hive
beeline>
beeline> !connect jdbc:hive2://localhost:10000
Connecting to jdbc:hive2://localhost:10000
Enter username for jdbc:hive2://localhost:10000: hduser
Enter password for jdbc:hive2://localhost:10000: *****
Connected to: Apache Hive (version 2.1.0)
Driver: Hive JDBC (version 3.1.3)
Transaction isolation: TRANSACTION_REPEATABLE_READ
0: jdbc:hive2://localhost:10000> show databases;
+----------------+
| database_name  |
```

```
+------------------+
| default          |
+------------------+
```

此处输入的用户名及密码是在配置文件 hive-site.xml 中设置的用户名和密码，这里分别是 hduser、hduser。操作界面如图 2-47 所示。

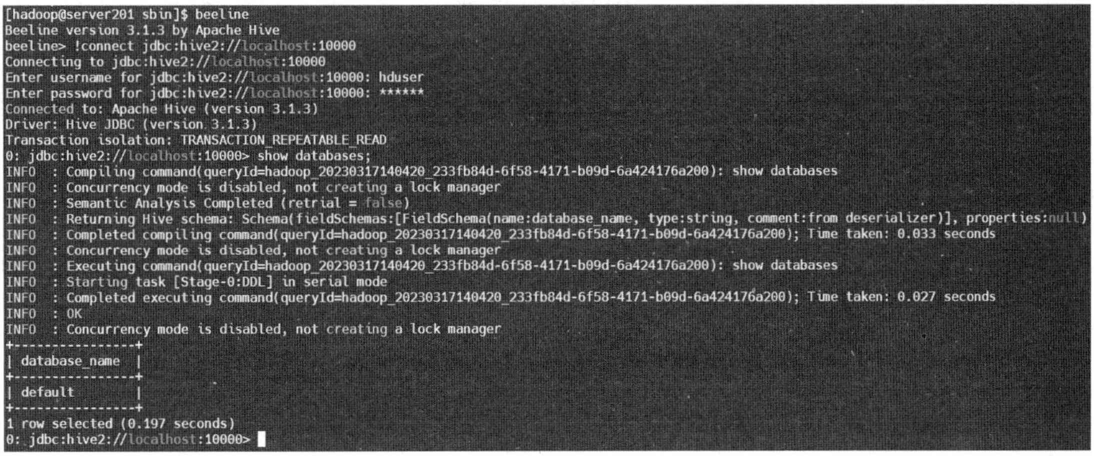

```
[hadoop@server201 sbin]$ beeline
Beeline version 3.1.3 by Apache Hive
beeline> !connect jdbc:hive2://localhost:10000
Connecting to jdbc:hive2://localhost:10000
Enter username for jdbc:hive2://localhost:10000: hduser
Enter password for jdbc:hive2://localhost:10000: ******
Connected to: Apache Hive (version 3.1.3)
Driver: Hive JDBC (version 3.1.3)
Transaction isolation: TRANSACTION_REPEATABLE_READ
0: jdbc:hive2://localhost:10000> show databases;
INFO  : Compiling command(queryId=hadoop_20230317140420_233fb84d-6f58-4171-b09d-6a424176a200): show databases
INFO  : Concurrency mode is disabled, not creating a lock manager
INFO  : Semantic Analysis Completed (retrial = false)
INFO  : Returning Hive schema: Schema(fieldSchemas:[FieldSchema(name:database_name, type:string, comment:from deserializer)], properties:null)
INFO  : Completed compiling command(queryId=hadoop_20230317140420_233fb84d-6f58-4171-b09d-6a424176a200); Time taken: 0.033 seconds
INFO  : Concurrency mode is disabled, not creating a lock manager
INFO  : Executing command(queryId=hadoop_20230317140420_233fb84d-6f58-4171-b09d-6a424176a200): show databases
INFO  : Starting task [Stage-0:DDL] in serial mode
INFO  : Completed executing command(queryId=hadoop_20230317140420_233fb84d-6f58-4171-b09d-6a424176a200); Time taken: 0.027 seconds
INFO  : OK
INFO  : Concurrency mode is disabled, not creating a lock manager
+----------------+
| database_name  |
+----------------+
| default        |
+----------------+
1 row selected (0.197 seconds)
0: jdbc:hive2://localhost:10000>
```

图 2-47　操作界面

第3章

Hive语法基础

在Hive中，数据类型、数据定义语言、数据操纵语言、函数等是英文字母大小写不敏感的。本书在提到这些具体的语句时，会针对不同情形混合使用英文大小写字母。

3.1 数据类型列表

Hive的基本数据类型如表3-1所示。

表3-1 基本数据类型及长度

类 型	长 度
tinyint	1byte，有符号整数
smallint	2byte
int	4byte
bigint	8byte
boolean	True|False
float	浮点
double	双精度浮点
string	字符串
timestamp	整数类型，或字符串
binary	字节数组类型

在创建Hive表时，必须指定表字段的数据类型。

Hive中的数据类型分为基本数据类型和复杂数据类型：基本数据类型包括数值类型、布尔类型、字符串类型、时间戳类型等，复杂数据类型包括数组（array）类型、映射（map）类型和结构体（struct）类型等。

Hive基本数据类型中的string类型相当于MySQL数据库中的varchar类型，该数据类型是一个可变长的字符串，理论上可以存储 2GB 的字符。

Hive 中涉及的日期时间有两种类型：

- 第一种是 date 类型，date 类型的数据就是通常所说的日期，常用“年、月、日”来表示一个具体的日期。date 类型的格式为 YYYY-MM-DD，其中 YYYY 表示年，MM 表示月，DD 表示日。Hive 中的 date 类型只包括年、月、日，不包括时、分、秒。
- 第二种是 timestamp 类型，它是与时区无关的类型，也就是说，各个时区、各个地方所表示的值是相等的，是一个从 UNIX 时代开始的时间偏移量。当前使用的时间戳偏移量都是 10 位整数，如果遇到 13 位的时间戳，则表示毫秒数。如果 timestamp 为浮点数，则表示精确到纳秒，小数点后保留 9 位。

在 Hive 中提供的 timestamp 可转换为日期，其格式为 YYYY-MM-DD HH:MM:SS。

3.2 集合类型

Hive 有 3 种复杂数据类型，如表 3-2 所示，包括数组、映射和结构体。array 和 map 与 Java 语言中的 array 和 map 类似，struct 与 C 语言中的 struct 类似。

表3-2 Hive的复杂数据类型

类型	示例
struct	对象或是结构体 hive> select struct('Jack','Mary'); OK _c0 {"col1":"Jack","col2":"Mary"}
map	map(key0, value0, key1, value1...) hive> select map('name','Jack','age',34); OK _c0 {"name":"Jack","age":"34"}
array	hive> select array(1,2,3); OK _c0 [1,2,3]

array 类型声明格式为 array<data_type>，表示相同数据类型的数据所构成的集合。array 元素的访问通过从 0 开始的下标实现，例如 array[1]访问的是第 2 个数组元素。

map 类型通过 map<key, value>来声明，key 只能是基本数据类型，value 可以是任意数据类型。map 元素的访问使用[]，例如 map['key1']。

struct 类型封装一组有名字的字段，可以包含不同数据类型的元素，其类型可以是任意的基本数据类型。struct 类型更灵活，可以存储多种数据类型的数据。struct 元素的访问使用点运算符。

下面演示创建一个使用集合类型的表：

```
hive> create table tb01(
    >  name string,
    >  arr array<string>,
    >  mm map<string,string>,
    >  ss struct<a:string,b:string>
    > );
OK
Time taken: 0.106 seconds
```

3.2.1 array 测试

array 是一种数组类型，存放的是相同类型的数据。

（1）测试 array：

```
hive> create table tb02(id int,names array<string>)
    > row format delimited fields terminated by '\t'
    > collection items terminated by '/';
OK
Time taken: 0.183 seconds
```

注意：代码中的 collection iterms terminated by '/'即指定/为数组分组的分隔符号。

（2）准备数据：

```
100     Jack/33
200     Mary/44
```

（3）导入数据，可以使用 overwrite 覆盖表中的原有数据：

```
hive> load data local inpath '/home/hadoop/arr.txt' overwrite into table tb02;
Loading data to table default.tb02
Table default.tb02 stats: [numFiles=1, numRows=0, totalSize=24, rawDataSize=0]
OK
Time taken: 0.413 seconds
hive> select * from tb02;
OK
tb02.id tb02.names
100 ["Jack","33"]
200 ["Mary","44"]
Time taken: 0.113 seconds, Fetched: 2 row(s)
```

从上面代码可以看出，已经通过/对 array 进行了分组处理。

（4）也可以使用下标查找数组中的数据：

```
hive> select id,names[0],names[1] from tb02;
OK
id  c1      c2
100 Jack    33
200 Mary    44
```

3.2.2　map 测试

若在 Hive 表中定义 map 类型数据，则 map 只能以键—值对的方式定义一批数据的数据类型。与 struct 相比，map 对每个字段的类型定义没有那么灵活。

（1）创建表：

```
hive> create table tb04(id int,info map<STRING,STRING>)
    > row format delimited
    > fields terminated by '\t'
    > collection items terminated by ','
    > map keys terminated by ':';
OK
Time taken: 0.107 seconds
```

（2）准备数据：

```
name:Jack,age:33,addr:SDJN
Name:Mary,age:55,addr:北京
```

（3）加载数据并查询：

```
hive> load data local inpath "/home/hadoop/map.txt' overwrite into table tb04;
Loading data to table default.tb04
Table default.tb04 stats:[numFiles=1,numRows=0,totalSize=60,rawDataSize=0]
OK
Time taken:0.417 seconds
 hive> select * from tb04;
OK
tb04.id tb04.info
{"name":"Jack","age":"33","addr":"SDJN"}1
{"name"："Mary", "age"："55", "addr"："北京"}2
Time taken:0.089 seconds,Fetched:2 row(s)
```

（4）其他更多查询：

```
hive> select sum(info['age']) from tb04;
```

3.2.3　struct 测试

Hive 数据类型的 struct 结构体与 Java 中类的结构非常相似。

（1）创建一张表，并指定分隔符号：

```
hive> create table tb03(id int,info struct<name:string,age:int>)
> row format delimited fields terminated by ','
>collection items terminated by':';
OK
Time taken:0.119 seconds
```

（2）准备数据：

```
1,Jack:23
2,Mary:44
100,Alex:45
12,Smith:66
```

```
3,张三：99
```

（3）导入数据，然后查询：

```
hive> load data local inpath '/home/hadoop/struct.txt' overwrite into table tb03;
Loading data to table default.tb03
Table default.tbo3 stats: [numFiles=1, numRows=0, totalSize=56,
rawDataSizem0]OK
Time taken:0.381 seconds
 hive> select * from tb03;
OK
tb03.id     tb03.info
{"name":"Jack","age":23}1
{"name":"Mary","age":44}2
{"name":"Alex","age":45}100 12
{"name":"Smith","age":66}{"name"："张三", "age"：99}0
Time taken: 0.111 seconds, Fetched: 5 row(s)
```

（4）使用字段名查询，在查询时，可以通过“结构体.列名”的格式指定列：

```
hive> select id,info.name,info.age from tb03;
OK
id  name    age
1   Jack    23
2   Mary    44
100 Alex    45
12  Smith   66
3   张三    99
```

3.3 数据类型转换

Hive 的基本数据类型可以根据需要进行类型转换，例如，tinyint 类型的数据与 int 类型的数据相加，则会将 tinyint 类型的数据隐式地转换成 int 类型的数据，然后与 int 类型的数据做加法，这个操作类似于 Java 的自动类型转换。数据类型转换分为隐式数据类型转换和强制数据类型转换。

- 任何整数类型都可以隐式地转换为一个范围更广的数据类型，如 tinyint 类型可以转换成 int 类型，int 类型可以转换成 bigint 类型。
- 所有整数类型及 float 类型、string 类型都可以隐式地转换成 double 类型。
- tinyint 类型、smallint 类型和 int 类型都可以隐式地转换为 float 类型。
- boolean 类型不可以转换为任何其他类型。
- timestamp 类型和 data 类型可以隐式地转换成文本类型。

有些情况需要数据类型的强制转换，数据类型强制转换的语法格式为：

```
CAST(expr AS <type>)
```

例如，cast('10' as int)将把字符串'10'转换成整数 10。

Hive 可以在 timestamp 类型、date 类型和字符串类型之间进行强制转换，例如：

```
cast(date as timestamp)
```

```
cast(timestamp as date)
cast(string as date)
cast(date as string)
```

如果数据类型强制转换失败，如执行 cast('X' as int)，则表达式返回空值 Null。

3.4 运 算 符

Hive 有 4 种类型的运算符：算术运算符、比较运算符、逻辑运算符和复杂运算符。这些运算符实际上是由 Hive 的内置函数实现的。运算符和操作数构成的表达式总能运算得到特定的结果。操作数可以是表的列名，即字段名。

算术运算符支持操作数的各种常见的算术运算，返回数值类型。表 3-3 描述了 Hive 常用的算术运算符。其中 A、B 是表的列名（即字段名），均为数值类型。

表3-3　算术运算符

算术运算符表达式	描　述
A+B	A 和 B 相加
A-B	A 减去 B
A*B	A 和 B 相乘
A/B	A 除以 B
A%B	A 对 B 取余

比较运算符也叫关系运算符，用来比较两个操作数。比较运算符表达式的返回值为 True、False 或 Null。表 3-4 描述了 Hive 常用的比较运算符。

表3-4　比较运算符

比较运算符表达式	支持的数据类型	描　述
A=B	基本数据类型	若 A 等于 B，则返回 True，否则返回 False
A<=>B	基本数据类型	若 A 和 B 都为 Null，则返回 True；若任一为 Null，则返回 Null；其他情况同等号（=）
A<>B, A!=B	基本数据类型	若 A 不等于 B，则返回 True，否则返回 False；A 或 B 为 Null，则返回 Null
A<B	基本数据类型	若 A 小于 B，则返回 True，否则返回 False；若 A 或 B 为 Null，则返回 Null
A<=B	基本数据类型	若 A 小于或等于 B，则返回 True，否则返回 False；若 A 或 B 为 Null，则返回 Null
A>B	基本数据类型	若 A 大于 B，则返回 True，否则返回 False；若 A 或 B 为 Null，则返回 Null
A>=B	基本数据类型	若 A 大于或等于 B，则返回 True，否则返回 False；若 A 或 B 为 Null，则返回 Null

逻辑运算符与字段构成逻辑表达式，并返回 True 或 False。Hive 常用的逻辑运算符如表 3-5 所示。

表3-5 逻辑运算符

逻辑运算符表达式	含 义
A And B	逻辑并
A Or B	逻辑或
A Not B	逻辑否

3.5 Hive 表存储格式

Hive 在创建表时需要指明该表的存储格式。Hive 支持的表存储格式主要有 TextFile、SequenceFile、ORC、Parquet，其中 TextFile 为默认格式。

Hive 表的存储方式如图 3-1 所示，其中图 3-1（a）所示为逻辑表，图 3-1（b）所示为行式存储，图 3-1（c）所示为列式存储。

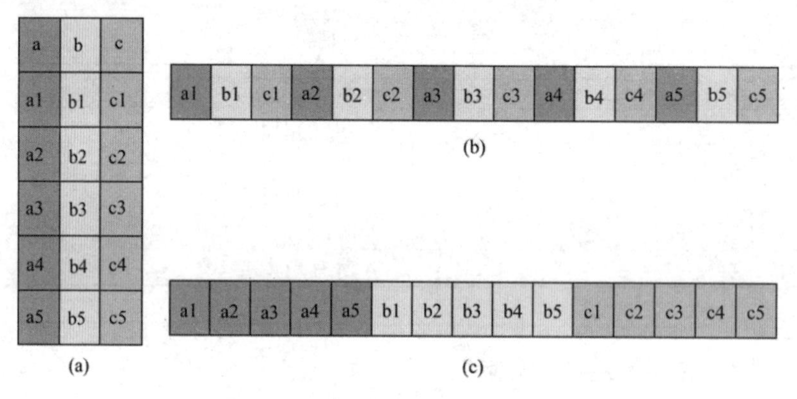

图 3-1 Hive 表的存储方式

当查询满足条件的一整行数据时，行式存储只需要找到其中一个值，其余的值都在相邻地方；列式存储则需要去每个聚集的字段中找到对应的每个字段的值，所以行式存储查询的速度更快。

列式存储中，每个字段的数据聚集存储，在查询只需要少数几个字段时，能大大减少读取的数据量；因为每个字段的数据类型一定是相同的，所以列式存储可以有针对性地设计更好的压缩算法。

TextFile 和 SequenceFile 的存储格式是基于行式存储的，ORC（Optimized Row Columnar）和 Parquet 的存储格式是基于列式存储的。

（1）TextFile 格式是 Hive 在创建表时默认的存储格式，数据不压缩，磁盘开销大，数据解析开销也较大。可结合 Gzip、Bzip2 压缩方式使用，但使用 Gzip 压缩方式时，Hive 不会对数据进行分割，因此无法对数据进行并行计算。

（2）SequenceFile 格式是为 Hadoop 存储二进制数形式的<key,value>而设计的一种文件格式。在存储结构上，SequenceFile 主要由一个头部（Header）及其后面跟着的多条记录（Record）组成。

- Header 主要包含存储压缩算法、用户自定义元数据等信息。此外，还包含一些同步标识，用于快速定位到记录的边界。
- 每条记录以键－值对的方式进行存储，用来表示它的字符数组可以一次性解析为记录的长度、key 的长度、key 值和 value 值，并且 value 值的结构取决于该记录是否被压缩。

（3）ORC 格式是 Hive 0.11 版本引入的表存储格式。每个 ORC 文件由一个或多个 Stripe 组成，每个 Stripe 大小为 250MB。每个 Stripe 由 3 部分组成，分别是 Index Data、Row Data 和 Stripe Footer。默认为每隔 1 万行做一个索引，该索引只记录某行的各字段在 Row Data 中的偏移量（Offset）。ORC 存储的是具体的数据，先取部分行，然后对这些行按列进行存储；对每个列进行编码，分成多个 Stream 来存储。对于 Stripe Footer，ORC 存储的是各个 Stream 的类型、长度等信息。

每个 ORC 文件还有一个 File Footer，这里面存储的是每个 Stripe 的行数、每列的数据类型信息等。每个 ORC 文件的尾部是一个 PostScript，这里面记录了整个文件的压缩类型及 File Footer 的长度信息等。在读取文件时，首先会 Seek（寻找）到文件尾部读取 PostScript，从里面解析得到 File Footer 长度，然后读 File Footer，从里面解析得到各个 Stripe 信息，再读各个 Stripe，即从后往前读。

（4）Parquet 格式文件是以二进制数形式存储的，因此是不可以直接读取的，文件中包含该文件的数据和元数据，因所以 Parquet 格式文件是自解析的。

通常情况下，在存储 Parquet 数据时会按照 Block（块）的大小设置行组的大小。一般情况下，每个 Map 任务处理数据的最小单位是一个 Block，这样可以把每个行组由一个 Map 任务处理，增加了任务执行并行度。

一个 Parquet 格式文件可以存储多个行组，文件的首位都是该文件的 Magic Code，用于校验是否是一个 Parquet 文件。Footer Length 记录了文件元数据的大小，通过该值和文件长度可以计算出元数据的偏移量。文件的元数据中包括每个行组的元数据信息和该文件存储数据的 Schema 信息。

3.6 Hive 的其他操作命令

Hive 的 CLI 命令行模式是最常用的模式。我们知道，启动 Hive 命令行模式有两种方式：在 Linux 终端直接输入 hadoop@server201:/usr/local/hive$ bin/hive 或 hadoop@server201:/usr/local /hive$ bin/hive --service cli。

Hive 的 CLI 提供了执行 HQL 语句、设置参数等功能。在命令行模式下执行 hive -help 命令，可以查看 CLI 的参数选项，这些参数选项是 Hive 常用的交互命令：

```
hadoop@server201:/usr/local/hive$ bin/hive -help
usage: hive
 -d,--define <key=value>          Variable subsitution to apply to hive
                                  commands. e.g. -d A=B or --define A=B
    --database <databasename>     Specify the database to use
 -e <quoted-query-string>         SQL from command line
 -f <filename>                    SQL from files
 -H,--help                        Print help information
    --hiveconf <property=value>   Use value for given property
    --hivevar <key=value>         Variable subsitution to apply to hive
                                  commands. e.g. --hivevar A=B
 -i <filename>                    Initialization SQL file
 -S,--silent                      Silent mode in interactive shell
```

```
  -v,--verbose                        Verbose mode (echo executed SQL to the
                                      console)
```

Hive 中的数据保存在 HDFS 中，因此可以在 Hive 命令行中直接执行 HDFS 的操作命令来查看 Hadoop 文件系统上的信息。在 Hive 中也可以执行 Linux 操作系统的命令。

（1）在 Hive CLI 命令窗口中查看 HDFS：

```
hive(default)>dfs -ls /user/hive/warehouse;
```

（2）在 Hive CLI 命令窗口中查看 HDFS 内容数据：

```
hive(default)>dfs -cat /user/hive/warehouse/test/test.txt;
```

（3）在 Hive CLI 命令窗口中查看本地文件系统：

```
hive(default)>! ls /opt/datas;
```

（4）查看在 Hive 中输入的所有历史命令：

进入当前用户的根目录/root 或/home/hadoop，查看.hivehistory 文件：

```
[hadoop@server201 ~]$ cat .hivehistory
```

（5）执行 SQL 脚本文件：

```
hive(default)>source SQL 脚本文件;
```

（6）清屏：

```
hive(default)>ctrl+L 或者!clear;
```

（7）退出 Hive 的两种方法：

```
hive(default)>exit;
hive(default)>quit;
```

3.7 Hive 分析 Tomcat 日志案例

本节介绍 Hive 的一个简单的应用案例——使用 Hive 分析 Tomcat 访问日志，日志信息如图 3-2 所示。

```
127.0.0.1 - - [02/Aug/2017:09:33:39 +0800] "GET / HTTP/1.1" 404 1070
0:0:0:0:0:0:0:1 - - [02/Aug/2017:09:33:40 +0800] "GET /20170731/ HTTP/1.1" 304 -
0:0:0:0:0:0:0:1 - - [02/Aug/2017:09:33:40 +0800] "GET /20170731/first HTTP/1.1" 200 1342
0:0:0:0:0:0:0:1 - - [02/Aug/2017:09:33:44 +0800] "GET /20170731/one HTTP/1.1" 200 123
127.0.0.1 - - [02/Aug/2017:09:35:19 +0800] "GET / HTTP/1.1" 404 1070
0:0:0:0:0:0:0:1 - - [02/Aug/2017:09:35:21 +0800] "GET /20170731/one HTTP/1.1" 200 287
127.0.0.1 - - [02/Aug/2017:09:44:20 +0800] "GET / HTTP/1.1" 404 1070
0:0:0:0:0:0:0:1 - - [02/Aug/2017:09:44:20 +0800] "GET /20170731/one HTTP/1.1" 200 1453
```

图 3-2　日志信息

步骤 01 创建表（注意，日志中包含的“-（短画线）”在内的数据也应该当作列来处理，后期将这些数据删除即可）：

```
hive> create table tb06(
    >  ip string,a string,b string,dt string,zone string,method string,url
string,protocol string,status string,account bigint)
    > row format delimited
    > fields terminated by ' ';
OK
Time taken: 0.536 seconds
```

步骤 02 导入数据：

```
hive> load data local inpath '/home/hadoop/access.log' into table tb06;
Loading data to table default.tb06
Table default.tb06 stats: [numFiles=1, totalSize=80325]
OK
Time taken: 1.554 seconds
```

步骤 03 查询前几行数据查看效果：

```
hive> select * from tb06 limit 4;
OK
tb06.ip tb06.a  tb06.b  tb06.dt tb06.zone   tb06.method tb06.url
tb06.protocol   tb06.status tb06.account
127.0.0.1   -   -   [02/Aug/2017:09:33:39   +0800]  "GET    /   HTTP/1.1"
404 1070
0:0:0:0:0:0:0:1 -   -   [02/Aug/2017:09:33:40   +0800]  "GET    /20170731/
HTTP/1.1"   304 NULL
0:0:0:0:0:0:0:1 -   -   [02/Aug/2017:09:33:40   +0800]  "GET
/20170731/first HTTP/1.1"   200 1342
0:0:0:0:0:0:0:1 -   -   [02/Aug/2017:09:33:44   +0800]  "GET    /20170731/one
HTTP/1.1"   200 123
Time taken: 0.206 seconds, Fetched: 4 row(s)
```

步骤 04 统计出现的 IP 的个数：

```
hive> create table tb07 as select count(1),ip from tb06 group by ip;
Query ID = hadoop_20180706154311_c0f908f2-742f-4213-9f72-21d1f37da188
Total jobs = 1
Launching Job 1 out of 1
```

步骤 05 执行完成以后，查看结果：

```
hive> select * from tb07;
OK
tb07.c0 tb07.ip
174     0:0:0:0:0:0:0:1
87      127.0.0.1
6       192.168.0.100
...
```

上面结果中，“…”表示后面的数据都省略。

步骤 06 由于_c0 是 Hive 自动生成的列名，因此在这里修改它的名称：

```
Hive > alter table tb07 change '_c0' cnt bigint;
```

修改以后，再查看表结构：

```
hive> desc tb07;
```

```
OK
col_name    data_type   comment
cnt         bigint
ip          string
Time taken: 0.131 seconds, Fetched: 2 row(s)
```

步骤07 分析访问量统计：

```
hive> select ip,sum(account) as sa from tb06 group by ip;
Query ID = hadoop_20180706173118_2c7f8b72-80dc-4c09-84c1-88a9d3eb1387
Total jobs = 1
```

结果如下：

```
ip                  sa
0:0:0:0:0:0:0:1     20106229
127.0.0.1           35342
192.168.0.100       3749
192.168.0.17        534
192.168.0.21        2146
192.168.0.22        3090
192.168.0.24        3429
192.168.0.25        4536
192.168.0.31        1577
192.168.0.50        3563
```

第 4 章

Hive 数据定义

本章将主要介绍 Hive 数据仓库的增、删、改、查和表的增、删、改、查操作，这些是 Hive 数据仓库的重点内容之一。Hive 默认不支持插入、修改或是删除数据，也不支持事务。

4.1 数据库的增删改查

4.1.1 在默认位置创建数据库

Hive 中的数据库仅是一个目录或是命名空间，它的功能是：

（1）可以避免表名冲突。

（2）通常使用数据库产生不同的表或是表的逻辑组。

创建数据仓库的语法格式如下：

```
CREATE DATABASE [IF NOT EXISTS] <database name> LOCATION <dir>;
```

IF NOT EXISTS 是一个可选子句，通知用户如果该数据仓库不存在，就创建数据仓库，否则报错（有错误信息提示）。

所创建的数据仓库在 HDFS 中默认的存储路径是/user/hive/warehouse/，也可以通过关键字 LOCATION 来指定数据仓库在 HDFS 中存放的位置。

创建一个名为 hivedwh 的数据仓库，并存放在默认位置：

```
hive(default)>create database hivedwh;
```

创建 hivedwh 数据仓库后，可以使用浏览器直观地浏览 Hadoop 集群的 HDFS。注意，IP 地址为本地 Linux 系统的 IP 地址，端口号为 50070。详细信息如图 4-1 所示，可以看到所创建的数据仓库 hivedwh 实际上对应 HDFS 中的一个目录，并且自动加上了扩展名.db。

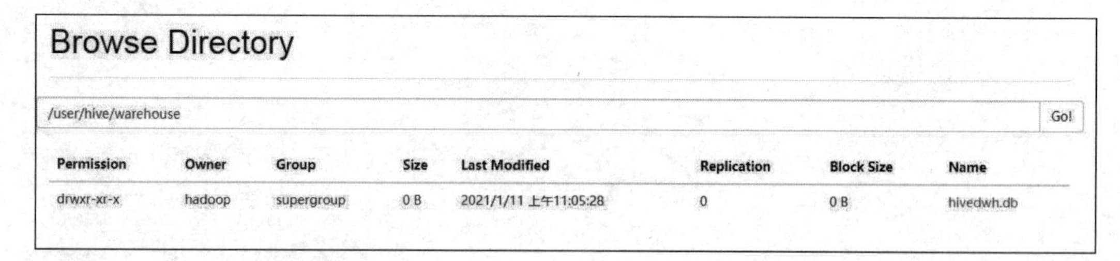

图 4-1　详细信息

创建一个数据仓库，并存放在 HDFS 的根目录下：

```
hive(default)>create database if not exists hivedwh2
location '/hivedwh2.db';
```

4.1.2　指定目录创建数据库

（1）创建一个目录：

```
hive> dfs -mkdir /db02 ;
hive> dfs -ls /db02;
```

（2）创建一个数据库，通过 location 指定它的目录为刚才创建的目录：

```
hive> create database db02
    > location '/db02';
OK
Time taken: 0.081 seconds
```

（3）使用这个数据库，并在这个数据库下创建一张表 tb01：

```
hive> use db02;
OK
Time taken: 0.071 seconds
hive> create table tb01(id int,name string)
    > row format delimited
    > fields terminated by '\t';
OK
Time taken: 0.286 seconds
```

（4）查看目录结构：

```
hive> dfs -ls -R /db02;
drwxr-xr-x   - hadoop supergroup          0 2018-07-06 22:27 /db02/tb01
```

（5）加载数据进去：

```
hive> load data local inpath '${env:HOME}/tb01.txt' into table tb01;
Loading data to table db02.tb01
Table db02.tb01 stats: [numFiles=1, totalSize=45]
OK
Time taken: 1.513 seconds
```

（6）查看里面的数据：

```
hive> dfs -ls -R /db02;
drwxr-xr-x   - hadoop supergroup          0 2018-07-06 22:30 /db02/tb01
```

```
-rwxr-xr-x   1 hadoop supergroup          45 2018-07-06 22:30 /db02/tb01/tb01.txt
```

（7）查询验证一下：

```
hive> select * from tb01;
OK
1    Jack
2    Mary
3    Rose
43   Alex
45   Rose
42   Uber
Time taken: 0.158 seconds, Fetched: 6 row(s)
```

4.1.3　显示当前使用的数据库

输入如下命令显示当前使用的数据库：

```
hive> set hive.cli.print.current.db=true;
hive (db02)>
```

设置完成以后，就会显示数据库。

4.1.4　删除数据库

输入如下命令删除数据库：

```
    hive (default)> drop database db02;
    FAILED: Execution Error, return code 1 from
org.apache.hadoop.hive.ql.exec.DDLTask. InvalidOperationException(message:
Database db02 is not empty. One or more tables exist.)
```

以上执行结果报出异常，说明数据库 db02 已存在，因此删除一个非空的数据库，将会返回一个异常。

可以使用 cascade 直接删除数据库里面的表：

```
hive (default)> drop database db02 cascade;
OK
Time taken: 1.099 seconds
```

描述更加详细的信息：

```
hive (default)> desc formatted tb07;
```

4.2　创建内部表

在 Hive 中，表都在 HDFS 的相应目录中存储数据。目录的名称是在创建表时自动创建并以表名来命名的，表中的数据都保存在该目录中，而且数据以文件的形式存储在 HDFS 中。表的元数据会存储在数据库中，如 Derby 数据库或 MySQL 数据库。

创建表的语法如下：

```
CREATE [EXTERNAL] TABLE [IF NOT EXISTS] table_name
(col_name data_type [COMMENT col_comment], ...)
[COMMENT table_comment]
[PARTITIONED BY (col_name data_type [COMMENT col_comment], ...)]
[CLUSTERED BY (col_name, col_name, ...) INTO num_buckets BUCKETS]
[SORTED BY (col_name [ASC|DESC], ...)]
[ROW FORMAT DELIMITED row_format]
[STORED AS file_format]
[LOCATION hdfs_path];
```

参数说明：

（1）CREATE TABLE：创建一个名字为 table_name 的表。如果该表已经存在，则抛出异常；可以用 IF NOT EXISTS 关键字来忽略异常。

（2）EXTERNAL：使用此关键字可以创建一个外部表，在建表的同时指定实际表数据的存储路径（LOCATION）。创建 Hive 内部表时，会将数据移动到数据仓库指定的路径；创建 Hive 外部表时，仅记录数据所在的路径，不对数据的位置做任何改变。在删除内部表时，内部表的元数据和数据会被一起删除；在删除外部表时，只删除外部表的元数据，但不删除数据。

（3）(col_name data_type [COMMENT col_comment], ...)：创建表时要确定字段名及其数据类型，数据类型可以是基本数据类型，也可以是复杂数据类型。COMMENT 为表和字段添加注释描述信息。

（4）PARTITIONED BY：创建分区表。

（5）CLUSTERED BY：创建桶表。

（6）SORTED BY：排序。

（7）ROW FORMAT DELIMITED：用于指定表中数据行和列的分隔符及复杂数据类型数据的分隔符。这些分隔符必须与表数据中的分隔符完全一致。

- [Fields Terminated By Char]，用于指定字段分隔符。
- [Collection Items Terminated By Char]，用于指定复杂数据类型 Map、Struct 和 Array 的数据分隔符。
- [Map Keys Terminated By Char]，用于指定 Map 中的 key 与 value 的分隔符。
- [Lines Terminated By Char]，用于指定行分隔符。

（8）STORED AS：指定表文件的存储格式，如 TextFile 格式、SequenceFile 格式、ORC 格式和 Parquet 格式等。如果文件数据是纯文本的，那么可以使用 TextFile 格式，这种格式是默认的表文件存储格式；如果数据需要压缩，则可以使用 SequenceFile 格式等。

（9）LOCATION：用于指定所创建表的数据在 HDFS 中的存储位置。

不带 EXTERNAL 关键字创建的表是管理表，有时也称为内部表。Hive 表是归属于某个数据仓库的，默认情况下 Hive 会将表存储在默认数据仓库中，也可以使用 Use 命令切换数据仓库，将所创建的表存储在切换后的数据仓库中。

删除内部表时，表的元数据和表数据文件同时被删除。

下面是内部表创建示例。

1. 需求

创建内部表 test，并将本地/opt/datas/test.txt 目录下的数据导入 Hive 的 test(id int, name string)表中。

2. 准备

在/opt/datas 目录下准备数据，创建 test.txt 文件并添加数据：

```
hadoop@server201:/opt/datas$ vim test.txt
101  Bill
102  Dennis
103  Doug
104  Linus
105  James
106  Steve
107  Paul
108  Ford
```

test.txt 文件中的数据以 Tab 键分隔。

3. Hive 实例操作

（1）启动 Hive：

```
hadoop@server201:/usr/local/hive$ bin/hive
```

（2）显示数据仓库：

```
hive(default)>show databases;
```

（3）切换到 hivedwh 数据仓库：

```
hive(default)>use hivedwh;
```

（4）显示 hivedwh 数据仓库中的表：

```
hive(hivedwh)>show tables;
```

（5）创建 test 表，并声明文件中数据的分隔符：

```
hive(hivedwh)>create table test(id int, name string)
row format delimited fields terminated by '\t';
```

（6）加载/opt/datas/test.txt 文件到 test 表中：

```
hive(hivedwh)>load data local inpath '/opt/datas/test.txt' into table test;
```

（7）Hive 查询结果：

```
hive(hivedwh)>select id,name from test;
OK
```

4.3　使用关键字 external 创建外部表

外部表可以加载外部的数据，本节介绍如何使用 external 关键字创建外部表。

4.3.1 指定现有目录

首先在 HDFS 上创建目录，并创建两个文本文件 tb01.txt 和 tb02.txt，创建目录的命令如下：

```
[hadoop@server201 ~]$hdfs dfs -mkdir /db
[hadoop@server201 ~]$hdfs dfs -put tb01.txt /db
[hadoop@server201 ~]$hdfs dfs -put tb02.txt /db
[hadoop@server201 ~]$hdfs dfs -ls -R /db
-rw-r--r--   1 hadoop supergroup        45 2018-07-06 22:54 /db/t01.txt
-rw-r--r--   1 hadoop supergroup        45 2018-07-06 22:54 /db/t02.txt
```

再创建外部表，并指定这个目录：

```
hive (db01)> create external table tb01(
           > id int,
           > name string)
           > row format delimited
           > fields terminated by '\t'   //指定文件分隔符
           > location '/db'              //指定 HDFS 目录
           > ;
OK
Time taken: 0.123 seconds
```

然后查询表中的数据，就可以获取到文件中的数据了：

```
hive (db01)> select * from tb01;
OK
1   Jack
2   Mary
3   Rose
43  Alex
45  Rose
42  Uber
1   Jack
2   Mary
3   Rose
43  Alex
45  Rose
42  Uber
Time taken: 0.147 seconds, Fetched: 12 row(s)
```

4.3.2 先创建表，再指定目录

下面我们先创建表，并为表创建分区，然后为分区表指定具体目录，不同分区对应不同的子目录。现在存在以下目录：

```
[hadoop@server201 ~]$ hdfs dfs -ls -R /db
drwxr-xr-x   - hadoop supergroup    0 2018-07-07 19:38 /db/tb01
drwxr-xr-x   - hadoop supergroup    0 2018-07-07 19:41 /db/tb02
```

（1）创建表并指定分区方案（注意，这里并没有使用 external 指定为外部表）：

```
hive (db01)> create table tb01(
           > id int,
           > name string)
```

```
            > partitioned by (year int,month int)
            > row format delimited
            > fields terminated by '\t';
    OK
    Time taken: 0.128 seconds
```

（2）指定第一个分区的值，并指定数据所在的外部区域：

```
    hive (db01)> alter table tb01 add partition(year=2018,month=1) location
'/db/tb01';
    OK
    Time taken: 0.199 seconds
```

通过上面的测试可以知道，创建的非外部表也可以加载外部的数据。

```
    hive (db01)> show partitions tb01;
    OK
    year=2018/month=1
    Time taken: 0.133 seconds, Fetched: 1 row(s)
```

4.3.3　显示某个表或某个分区的信息

使用 show 显示某个表或某个分区的信息，语法如下：

```
    SHOW TABLE EXTENDED[IN|FROM database_name] LIKE identifier_with_wildcards
[PARTITION (partition_desc)]
```

以下示例用于显示 tb01 表的详细信息：

```
    hive (db01)> show table extended like tb01; //执行命令
    OK
    tableName:tb01 表名
    owner:hadoop 拥有者
    location:hdfs://server201:8020/user/hive/warehouse/db01.db/tb01 存储位置
    inputformat:org.apache.hadoop.mapred.TextInputFormat 输入类型
    outputformat:org.apache.hadoop.hive.ql.io.HiveIgnoreKeyTextOutputFormat 输出
类型
    columns:struct columns { i32 id, string name} 字段信息，为结构体
    partitioned:true  是否分区
    partitionColumns:struct partition_columns { i32 year, i32 month} 分区信息，可以
看出都是结构体
    totalNumberFiles:1
    totalFileSize:55
    maxFileSize:55
    minFileSize:55
    lastAccessTime:1530962079409
    lastUpdateTime:1530963924748
    Time taken: 0.174 seconds, Fetched: 15 row(s)
```

以下示例显示某个分区的具体信息：

```
    hive (db01)> show table extended like tb01 partition(year=2018,month=1); 执行
命令
    OK
    tableName:tb01
    owner:hadoop
```

```
location:hdfs://server201:8020/db/tb01 数据存储的位置，这是重点要了解的信息，其他的同上
inputformat:org.apache.hadoop.mapred.TextInputFormat
outputformat:org.apache.hadoop.hive.ql.io.HiveIgnoreKeyTextOutputFormat
columns:struct columns { i32 id, string name}
partitioned:true
partitionColumns:struct partition_columns { i32 year, i32 month}
totalNumberFiles:1
totalFileSize:55
maxFileSize:55
minFileSize:55
lastAccessTime:1530962079409
lastUpdateTime:1530963924748
Time taken: 0.298 seconds, Fetched: 15 row(s)
```

4.4 创建分桶表

创建分桶表的步骤说明如下：

步骤01 先创建一个表，然后将数据都导入这个表中。

步骤02 再创建一个分桶的表。

步骤03 设置分桶变量为开：hive.enforce.bucketing=true。

步骤04 设置 reduce 个数与分桶的个数相同，mapreduce.job.reduces=n。

步骤05 将数据通过 insert into...select...distribute by...sort by...语句导入表中。

（1）准备数据：

```
ID      Name    出厂年   价格
C001    BMW     2017    40
C002    Buick   2009    18
C003    BMW     2008    36
C004    AUDI    2007    50
```

（2）创建表：

```
create table cars(
  id string,
  name string,
  year int,
  price double
)
clustered by (id) into 3 buckets
row format delimited
fields terminated by '\t';
```

（3）使用 source 执行代码：

```
hive (db01)> source ${env:HOME}/cars.sql;
OK
Time taken: 0.236 seconds
```

（4）加载数据：

```
hive (db01)> load data local inpath '${env:HOME}/cars.txt' overwrite into table cars;
Loading data to table db01.cars
Table db01.cars stats: [numFiles=1, numRows=0, totalSize=71, rawDataSize=0]
OK
Time taken: 0.555 seconds
```

（5）设置强制分桶：

```
hive (db01)> set hive.enforce.bucketing=true;
hive (db01)> set hive.enforce.bucketing;
hive.enforce.bucketing=true
```

（6）设置Reduce的数量：

```
hive (db01)> set mapreduce.job.reduces=3;
```

（7）删除数据以后重新导入数据。

要删除数据，可以直接删除导入的数据文件：

```
hive (db01)> dfs -rm /user/hive/warehouse/db01.db/cars/*;
Deleted /user/hive/warehouse/db01.db/cars/cars.txt
hive (db01)> select * from cars;
OK
Time taken: 0.113 seconds
```

再重新导入数据：

```
 load data local inpath '${env:HOME}/cars.txt'
overwrite into table cars;
```

（8）再次查看里面的数据文件，还是只有一个，即没有分桶：

```
hive(db01)>dfs -ls /user/hive/warehouse/db01.db/cars;
Found 1 items
-rwxr-xr-x 1 hadoop supergroup   71 2018-07-10 15:22
/user/hive/warehouse/db01.db/cars/cars.txt
```

（9）现在再创建一个分桶的表，除了表名不一样，其他的都一样：

```
create table cars_buckets(
  id string,
  name string,
  year int,
  price double
)
clustered by (id) into 3 buckets
row format delimited
fields terminated by '\t'
```

（10）把表cars的数据导入上面的表cars_buckets中。

因为使用了distribute by且又指定了Reduce的个数，所以会分别输出到3个文件中：

```
hive (db01)> insert into table cars_buckets
> select id,name,year,price from cars
```

```
    >distribute by(id) sort by(id);
    Query ID= hadoop_20180710153043295cc076-740c-436b-88a7-e808b7b261a7 Total jobs = 2
    Launching Job 1 out of 2
    Number of reduce tasks not specified. Defaulting to jobconf value of: 3 In order
to change the average load for a reducer(in bytes):
    set hive.exec.reducers.bytes.per.reducer=<number>In order to limit the maximum
number of reducers:
    set hive.exec.reducers.max=<number>
    In order to set a constant number of reducers:
    set mapreduce.job.reduces=<number>
    Starting Job = job_1531183134586_0015,Tracking URL =
http://server201:8088/proxy/application
    Kill Command = /app/hadoop-3.2.3/bin/hadoop job -kill job 1531183134596 0015
Hadoop job information for Stage-1: number of mappers: 1; oumber of reducers: 3
2018-07-10 15:30:54,133 Stage-1 map = 0%, reduce = 0%
    2018-07-10 15:31:01,698 Stage-1 map = 100%, reduce = 0%, Cumulative CPU 1.32 sec
```

（11）查看数据，已经保存到了不同的文件中：

```
    hive(db01)>dfs -ls /user/hive/warehouse/db01.db/cars_buckets;
    Found 3 items
    -rwxr-xr-x      1  wangjian supergroup   21 2018-07-10 15:32
/user/hive/warehouse/db01.db/cars buckets/000000_0
    -rwxr-xr-x      1  wangjian supergroup   19 2018-07-10 15:32
/user/hive/warehouse/db01.db/cars buckets/000001_0
    -rwxr-xr-x      1  wangjian supergroup   39 2018-07-10 15:32
/user/hive/warehouse/db01.db/cars buckets/000002_0
```

（12）分别查看 3 个文件里面的数据：

```
    hive(db0l)>dfs -cat /user/hive/warehouse/db01.db/cars buckets/00000_0;
    C002 Buick 2009  18.0
    hive(db0l)>dfs -cat /user/hive/warehouse/db01.db/cars buckets/00001_0;
    C003 BMW 2008  36.0
    hive(db0l)>dfs -cat /user/hive/warehouse/db01.db/cars buckets/00002_0;
    C004 AUDI  2017  50.0
    C001 BMW   2017  40.0
```

（13）查看 cars_buckets 表的 Buckets 信息：

```
     hive -e "use db01;desc formatted cars_buckets;"
    #存储信息
    SerDe Library:  org.apache.hadoop.hive.serde2.lazy.LazySimpleSerDe
    InputFormat:  org.apache.hadoop.mapred.TextInputFormat
    OutputFormat:  org.apache.hadoop.hive.ql.io.Hive Ignore KeyTextOutputFormat
    Compressed: No
    Num Buckets:3
    Bucket Columns:[id]
    Sort Columns:[]
    Storage Desc Params:
    field.delim \t
    serialization.format  \t
```

4.5 分 区 表

分区表具有以下作用：

- 使用 partitioned by(name string,other string)来指定分区策略。
- 分区可以显著提高查询的性能。
- 可以将 Hive 设置成严格模式，在查询时必须传递分区的查询条件。语句如下：

```
hive (db01)> set hive.mapred.mode;
hive.mapred.mode=nonstrict
hive (db01)> set hive.mapred.mode=strict;
hive (db01)> set hive.mapred.mode;
hive.mapred.mode=strict
hive (db01)>
```

4.5.1 创建和显示分区表

（1）首先创建一张有分区的表：

```
hive (db01)> create table tb02(
           > id int,
           > name string)
           > partitioned by (grade int)
           > row format delimited
           > fields terminated by '\t';
OK
Time taken: 0.153 seconds
```

创建以后，表里面还没有数据：

```
hive (db01)> dfs -ls -R /user/hive/warehouse/db01.db;
drwxr-xr-x   - hadoop supergroup          0 2018-07-06 23:13
/user/hive/warehouse/db01.db/tb02
```

（2）导入数据并指定分区：

```
hive (db01)> select * from tb02;
OK
tb02.id tb02.name   tb02.grade
1   Jack    1
2   Mary    1
3   Rose    1
43  Alex    1
45  Rose    1
42  Uber    1
Time taken: 0.139 seconds, Fetched: 6 row
```

不建议在没有分区条件的情况下进行查询。如果一张表已经创建分区，默认在没有分区条件的情况下进行查询会抛出异常：

```
hive> set hive.mapred.mode;
hive.mapred.mode=strict
```

上述 hive.mapred.mode 的值可为 strict 或 nostrict。当值为 strict 时，在查询时必须传递分区的条件，如果不传递，则抛出以下异常：

```
hive> select * from tb02;
FAILED: SemanticException [Error 10041]: No partition predicate found for Alias "tb02" Table "tb02"
```

因此必须传递分区条件：

```
hive> show partitions tb02;  //显示这个表的分区
OK
grade=1
grade=2
Time taken: 0.333 seconds, Fetched: 2 row(s)
hive> select * from tb02 where grade=1;  //根据分区查询
OK
1   Jack    1
2   Mary    1
3   Rose    1
43  Alex    1
45  Rose    1
42  Uber    1
Time taken: 0.579 seconds, Fetched: 6 row(s)
```

（3）再次创建一张分区表：

```
hive> create table tb03(
    > id int,
    > name string)
    > partitioned by (year int,month int)
    > row format delimited
    > fields terminated by '\t';
OK
Time taken: 0.383 seconds
```

（4）加载数据并指定分区数值，注意最后指定的分区的值：

```
hive> load data local inpath '/home/hadoop/tb01.txt' into table tb03 partition(ycar-2018,month=1);
Loading data to table db01.tb03 partition (year=2018, month=1)
Partition db01.tb03{year=2018, month=1} stats: [numFiles=1, numRows=0, totalSize=45, rawDataSize=0]
OK
Time taken: 1.558 seconds
```

4.5.2 增加、删除和修改分区

（1）增加分区，并指定数据。以下添加两个分区：

```
hive (db01)> alter table tb01 add
           > partition(year=2018,month=1) location '/db/tb01'
           > partition(year=2018,month=2) location '/db/tb02';
OK
Time taken: 0.227 seconds
```

查看分区：

```
hive (db01)> show partitions tb01;
OK
year=2018/month=1
year=2018/month=2
Time taken: 0.121 seconds, Fetched: 2 row(s)
```

（2）让分区指向新的地址。在官网上提示可通过 set location 移动分区，不过这个更像是让原来的分区指向一个新的地址。

```
hive (db01)> alter table tb01 partition(year=2018,month=1)
           > set location 'hdfs://server201:8020/db/tb03';
OK
Time taken: 0.405 seconds
hive (db01)> select * from tb01 where year=2018 and month =1;
OK
500 张三     2018    1
600 李四     2018    1
1   Jerry    2018    1
3   Mark     2018    1
501 Rack     2018    1
```

（3）删除分区：

```
hive (db01)> alter table tb01 drop if exists partition(year=2018,month=3);
Dropped the partition year=2018/month=3
OK
Time taken: 0.32 seconds
```

（4）修改列的名称：

```
hive (db01)> alter table tb01 change column id tid int;
OK
Time taken: 0.349 seconds
hive (db01)> desc tb01;
OK
tid                     int
name                    string
year                    int
month                   int
```

（5）增加新的列。以下示例添加两个新的列：

```
hive (db01)> alter table tb01 add columns(
           >   addr string,
           >   tel string
           > );
OK
Time taken: 0.139 seconds
```

修改以后添加新的数据：

```
hive (db01)> alter table tb01 add partition(year=2018,month=4) location
'/db/tb04';
OK
Time taken: 0.151 seconds
```

加载数据以后，再次查询：

```
hive (db01)> alter table tb01 add partition(year=2018,month=4) location
'/db/tb04';
OK
Time taken: 0.151 seconds
hive (db01)> select * from tb01 where month=4;
OK
1000    Jack    山东济南 1890987889  2018    4
1001    Mary    NY      1398767564  2018    4
2003    Alex    US      1908757463  2018    4
29      张三    中国     1897635475  2018    4
Time taken: 0.231 seconds, Fetched: 4 row(s)
```

之前的数据将显示为 NULL：

```
hive> select * from tb01 where month in(1,4);
OK
500     张三    NULL    NULL        2018    1
600     李四    NULL    NULL        2018    1
1       Jerry   NULL    NULL        2018    1
3       Mark    NULL    NULL        2018    1
501     Rack    NULL    NULL        2018    1
1000    Jack    济南     1890987889  2018    4
1001    Mary    NY      1398767564  2018    4
2003    Alex    US      1908757463  2018    4
29      张三    中国     1897635475  2018    4
```

4.6 显示某张表的详细信息

1. desc

可以使用“desc[ribe] tableName;”命令查看表的信息：

```
hive> desc tb03;
OK
id                  int
name                string
year                int
month               int

# Partition Information
# col_name          data_type           comment     //分区信息

year                int
month               int
Time taken: 01137 seconds, Fetched:  10 row(s)
```

2. desc formatted

通过关键字 formatted 显示表的更加详细的信息：

```
hive> describe formatted tb03;
```

```
OK
# col_name              data_type               comment     //字段信息
id                      int
name                    string
# Partition Information
# col_name              data_type               comment     //分区信息
year                    int
month                   int
# Detailed Table Information
Database:               db01  //所在数据库的名称
Owner:                  hadoop
CreateTime:             Sat Jul 07 17:38:50 CST 2018
LastAccessTime:         UNKNOWN
Protect Mode:           None
Retention:              0
Location:               hdfs://server201:8020/user/hive/warehouse/db01.db/tb03//
这是保存的位置
Table Type:             MANAGED_TABLE   //管理表
Table Parameters:
    transient_lastDdlTime   1530956330

# Storage Information
SerDe Library:          org.apache.hadoop.hive.serde2.lazy.LazySimpleSerDe
InputFormat:            org.apache.hadoop.mapred.TextInputFormat    //输入类型
OutputFormat:
org.apache.hadoop.hive.ql.io.HiveIgnoreKeyTextOutputFormat//输出类型
Compressed:             No
Num Buckets:            -1
Bucket Columns:         []
Sort Columns:           []
Storage Desc Params:
    field.delim         \t
    serialization.format    \t
Time taken: 0.153 seconds, Fetched: 34 row(s)
```

3. desc extended 非格式化信息

也可以使用 desc extended 显示一个非格式化信息：

```
hive> desc extended tb03;
OK
id                      int
name                    string
year                    int
month                   int
# Partition Information
# col_name              data_type               comment
year                    int
month                   int
Detailed Table Information Table(tableName:tb03, dbName:db01, owner:hadoop,
createTime:1530956330, lastAccessTime:0, retention:0,
sd:StorageDescriptor(cols:[FieldSchema(name:id, type:int, comment:null),
FieldSchema(name:name, type:string, comment:null), FieldSchema(name:year, type:int,
comment:null), FieldSchema(name:month, type:int, comment:null)],
location:hdfs://server201:8020/user/hive/warehouse/db01.db/tb03,
```

```
inputFormat:org.apache.hadoop.mapred.TextInputFormat,
outputFormat:org.apache.hadoop.hive.ql.io.HiveIgnoreKeyTextOutputFormat,
compressed:false, numBuckets:-1, serdeInfo:SerDeInfo(name:null,
serializationLib:org.apache.hadoop.hive.serde2.lazy.LazySimpleSerDe,
parameters:{field.delim=    , serialization.format=
    Time taken: 0.232 seconds, Fetched: 12 row(s)
```

4.7 指定输入输出都是 SequenceFile 类型

SequenceFile 文件是为 Hadoop 存储二进制形式的<key,value>而设计的一种平面文件（Flat File）。

把 SequenceFile 当作一个容器，将所有的文件都打包到 SequenceFile 类中可以高效地对小文件进行存储和处理。SequenceFile 文件并不按照其存储的 key 进行排序存储，SequenceFile 的内部类 Writer 提供了 append 功能。SequenceFile 中的 key 和 value 可以是任意类型的 Writable 或者是自定义的 Writable。

在存储结构上，SequenceFile 主要由一个 Header 及其后面跟着的多条 Record 组成。Header 主要包含了 key classname、value classname、存储压缩算法、用户自定义元数据等信息，此外，还包含了一些同步标识，用于快速定位到记录的边界。每条 Record 以键－值对的方式进行存储，用来表示它的字符数组可以一次解析成：记录的长度、key 的长度、key 值和 value 值，并且 value 值的结构取决于该记录是否被压缩。

通过在创建表时添加以下内容来指定输入输出是 SequenceFile：

```
STORED AS SEQUENCEFILE
INPUTFORMAT ''
OUTPUTFORMAT ''
```

4.8 关 于 视 图

视图（view）在 Hive 中的用法与在 SQL 中的用法相同。视图是一种虚表，是一个逻辑概念，可以跨越多张表，操作视图和操作表是完全一样的。视图建立在已有表的基础上，并不存储数据。从视图中查询出来的数据都来自视图所依赖的表，视图赖以建立的表称为基表。视图可以根据用户的需求来创建，也可以将任何结果集数据保存为一个视图。可以在视图上执行所有 DML 操作。视图可以简化复杂的查询。

4.8.1 使用视图降低查询的复杂度

视图可以允许一个查询，并像对待表一样对这个查询进行操作。视图是一个逻辑结构，因为它不像表那样存储数据。首先创建一个视图：

```
use db01;
create view v_studbook
as
select s.name as sname,b.name as bname
```

```
    from stud s inner join books b on s.id=b.sid;
```

然后就可以对视图进行一些操作，以下操作都可以成功：

```
    [hadoop@server201 hive]$ hive -e "use db01;select * from v_studbook"; 简单查询
    [hadoop@server201 hive]$ hive -e "use db01;select * from v_studbook where sname
like '%a%'"; 条件查询
    [hadoop@server201 hive]$ hive -e "use db01;select count(1) from v_studbook";
函数查询
```

4.8.2　查看视图的信息

视图也是表，所以 show tables 也会显示视图。

可以通过"desc 视图名称"命令来显示这个视图的基本信息，比如：

```
hive (db01)> desc v_studbook;
OK
sname                 string
bname                 string
Time taken: 0.301 seconds, Fetched: 2 row(s)
```

或使用 desc formatted view_name 命令来显示这个视图的详细信息：

```
hive (db01)> desc formatted v_studbook;
OK
# col_name                 data_type            comment
sname                      string
bname                      string
...
# View Information
View Original Text:        select s.name as sname,b.name as bname
from stud s inner join books b on s.id=b.sid
View Expanded Text:        select 's'.'name' as `sname','b'.'name' as 'bname'
from 'db01'.'stud' 's' inner join 'db01'.'books' 'b' on 's'.'id'='b'.'sid'
```

4.8.3　删除视图

删除视图的语法格式如下：

```
DROP VIEW view_name;
```

例如，删除视图 test_view：

```
hive(hivedwh)>drop view test_view;
```

第5章

Hive 数据操作

Hive 数据操作包含 Load、Insert、Update 和 Delete 等。需要注意的是，频繁的 Update 和 Delete 操作违背了 Hive 的初衷，一般情况下，需要减少 Update 和 Delete 操作的使用。

本章将主要介绍 Hive 数据仓库的数据装载、数据导入、数据导出的操作方法。其中 5.1 节介绍数据装载，5.2～5.4 节介绍数据导入，5.5 节介绍数据导出。

5.1 向表中装载数据

数据装载是指创建 Hive 表后向 Hive 表中加载或插入数据。

一般来说，在 Hive 创建表后，可使用 Load Data 语句向表中加载数据，也可使用 Insert、As Select、Location、Import 语句向表中插入数据。

从数据所在位置来看，有 4 种常见的数据导入方式：

- 从 Linux 本地文件系统中导入数据到 Hive 表。
- 将 HDFS 中的数据导入 Hive 表中。
- 从其他表中查询出相应的数据并导入 Hive 表中。
- 在创建表时从其他表中查询出相应数据并导入所创建的表中。

通过 Load 命令方式向 Hive 表中加载数据，Load 命令不会在加载数据时做任何转换工作，而是纯粹地把数据文件复制或移动到 Hive 表所在目录。语法格式如下：

```
LOAD DATA [LOCAL] INPATH 'filepath' [OVERWRITE] INTO TABLE table_name [PARTITION (partcol1=val1,…)];
```

参数说明：

（1）LOAD DATA：表示加载数据。

（2）LOCAL：表示从本地加载数据到 Hive 表，这种方式是对数据的复制过程；否则从 HDFS 加载数据到 Hive 表，这种方式是对数据的移动过程。

（3）INPATH：表示加载数据的目录为 filepath（会加载目录下的所有文件），filepath 也可以是一个文件。filepath 可以是相对路径，如 datas/test.txt；filepath 也可以是绝对路径，如/opt/datas/test.txt；filepath 还可以是完整的 URL，如 hdfs://localhost:9000/user/hive/warehouse/ test.txt。

（4）OVERWRITE：表示加载数据之前会先清空目标表中的数据内容，否则就是追加的方式。

（5）INTO TABLE：表示把数据加载到 table_name 表中。

（6）PARTITION：表示把数据加载到指定分区表中。

使用 Load Data 语句向表中加载数据的示例如下：

```
    hive (db01)> load data local inpath '/home/hadoop/tb04.txt'
             > overwrite into table tb01
             > partition(year=2018,month=4);
    Loading data to table db01.tb01 partition (year=2018, month=4)
    Partition db01.tb01{year=2018, month=4} stats: [numFiles=1, numRows=0,
totalSize=110, rawDataSize=0]
    OK
    Time taken: 0.918 seconds
```

显示所有分区：

```
    hive (db01)> show partitions tb01;
    OK
    year=2018/month=1
    year=2018/month=2
    year=2018/month=4
    Time taken: 0.117 seconds, Fetched: 3 row(s)
```

可以发现显示出了 year=2018/month=4 这个分区。显示表中该分区的信息：

```
    hive (db01)> show table extended like tb01 partition(year=2018,month=4);
    OK
    tableName:tb01
    owner:hadoop
    location:hdfs://server201:8020/user/hive/warehouse/db01.db/tb01/year=2018/m
onth=4
```

可以看到，该分区默认保存到 warehouse 目录下。

5.2 通过 Insert 向表中插入数据

SequenceFile、Parquet、ORC 格式的表不能直接从本地文件导入数据。数据要先导入 TextFile 格式的表中，然后再用 Insert 命令把数据从 TextFile 格式的表中导入 SequenceFile、Parquet、ORC 格式的表中。

通过查询语句使用 Insert 命令方式向 Hive 表中插入数据的基本语法格式如下：

```
    INSERT OVERWRITE|INTO TABLE table_name [PARTITION(partcol1=val1,…)] select_
statement;
```

参数说明：

（1）OVERWRITE：表示覆盖已经存在的数据。
（2）INTO：表示只是简单地插入数据，不考虑原始表的数据，直接追加到表中。
（3）select_statement：是可以针对一张表，也可以针对多张表的 Select 语句。

Insert 插入数据方式会自动启动 MapReduce 作业。
下面来看一个通过 Insert 向表中插入数据的示例。
在创建一张表时，直接从另一个查询中获取数据：

```
hive (db01)> create table tb02
           > as select * from tb01 where year=2018;
Query ID = hadoop_20180707220947_bc20fbfc-66f1-42d0-a05c-b2d59d8717a6
Total jobs = 3
```

创建一张表，然后使用 Insert 写入记录：

```
hive (db01)> create table tb03(
           > id int,
           > name string)
           > row format delimited
           > fields terminated by '\t';
OK
Time taken: 0.096 seconds
```

写入记录：

```
hive (db01)> insert into tb03
           > select tid,name from tb01 where year=2018;
```

查看里面的数据，可见是以\t 分隔的：

```
[hadoop@server201 ~]$ hdfs dfs -cat /user/hive/warehouse/db01.db/tb03/000000_0
500 张三
600 李四
1   Jerry
3   Mark
501 Rack
1   Jack
2   Mary
```

5.3 动态分区插入数据

动态分区属性配置表如表 5-1 所示。

表5-1 动态分区属性配置表

属性名称	默认值	描述
hive.exec.dynamic.partition	false	是否开启动态分区功能
hive.exec.dynamic.partition.mode	strict	设置为 nostrict 表示允许所有分区都是动态的

动态分区的字段必须出现在查询的最后部分。

首先创建一个文件 tb05.txt，内容如下：

```
101	Jack	2018	1
102	Mary	2018	2
103	Mark	2018	1
104	Rose	2018	2
```

然后导入数据并查询：

```
hive (db01)> load data local inpath '/home/hadoop/tb05.txt' into table tb04;
Loading data to table db01.tb04
Table db01.tb04 stats: [numFiles=1, totalSize=64]
OK
Time taken: 0.292 seconds
hive (db01)> select * from tb04;
OK
101 Jack	2018	1
102 Mary	2018	2
103 Mark	2018	1
104 Rose	2018	2
Time taken: 0.108 seconds, Fetched: 4 row(s)
```

再创建一张表 tb05：

```
hive (db01)> create table tb05(
           >  id int,
           > name string
           > )
           > partitioned by(year int,month int)
           > row format delimited
           > fields terminated by '\t';
OK
Time taken: 0.113 seconds
```

动态创建（以下动态创建会失败）：

```
hive (db01)> insert overwrite table tb05
           > partition (year,month)
           > select id,name,year,month from tb04;
FAILED: SemanticException [Error 10096]: Dynamic partition strict mode requires
at least one static partition column. To turn this off set
hive.exec.dynamic.partition.mode=nonstrict
```

最后一句是要求设置打开动态分配功能。

设置成可以动态分配：

```
hive (db01)> set hive.exec.dynamic.partition.mode=nonstrict;
```

再次执行动态创建就可以进行动态分区：

```
hive (db01)> insert overwrite table tb05
           > partition (year,month)
           > select id,name,year,month from tb04;
Query ID = hadoop_20180707223112_a6b77b3a-205e-4562-aa8b-22210b7e5403
Total jobs = 3
Launching Job 1 out of 3
```

查询分区 tb05：

```
hive (db01)> show partitions tb05;
OK
year=2018/month=1
year=2018/month=2
Time taken: 0.099 seconds, Fetched: 2 row(s)
```

5.4 创建表并插入数据

通过 create table tableName as select...语句可以直接从查询结果中创建一张表。默认新表中的字段采用查询语句中的字段名；如果使用了聚合函数并设置了别名，则使用别名作为新的字段；如果使用了聚合函数，则创建的字段名称为_c0、_c1 这样的列名，此时如果引用这些名称，则应该使用`_c0`。

注意：这个反引号是键盘上 Tab 键上面的那个引号，是执行语句的含义。

示例：

```
create table someTable
as
select id,name from otherTable;
```

使用 create...as...方式创建表的好处如下：

（1）create...as...语句中使用查询的结果创建和填充表。由 CTAS 创建的表是原子的，这意味着在填充所有查询结果之前，其他用户不会看到该表。因此，其他用户要么看到带有完整查询结果的表，要么根本看不到表。

（2）在 create...as...语句中有两个部分，select 部分可以是 HiveQL 支持的任何选择语句，create 部分创建表的结构来自 select 部分的字段名，并可以使用 SerDe 和存储格式等其他表属性创建指定的目标表，比如指定行列切分格式等属性。

另外，还有一种半自动化模式，即使用 create table like...语句。create table like 创建表的形式允许我们精确地复制现有表定义（不复制它的数据），它创建的表除了表名和源表不一样外，其余所有的细节都是一样的，但是不包含源表的数据。因此，这种 create table like 的形式非常适合对源表模式的复制。实际开发中使用得不是太多。

5.5 导出数据

数据导出是指将查询结果或者 Hadoop 集群的 HDFS 中的数据导出到本地文件系统或者 HDFS 中的其他目录下。

1. 语法

如果某个文件格式恰好是用户所需要的格式，那么直接使用 hdfs cp 命令即可，否则可以使用 insert [overwrite local] directory select ...命令。

2. 导出到某个目录

以下示例将会开启 MapReduce 应用：

```
hive (db01)> insert overwrite directory '/tb05'
           > select id,name,year,month from tb05;
```

查看内容：

```
hive (db01)> dfs -cat /tb05/*;
101Jack20181
103Mark20181
102Mary20182
104Rose20182
```

3. 使用 cat -A 查看分隔符

从上面结果可以看出字段值之间并没有分隔符号，这是因为对于“\001”这样的默认分隔符并不显示。可以使用 cat -A 显示所有分隔符信息：

```
[hadoop@server201 ~]$ hdfs dfs -cat /tb05/* | head -3 | cat -A
101^AJack^A2018^A1$
103^AMark^A2018^A1$
102^AMary^A2018^A2$
```

结果中已经显示了分隔符。

4. 指定导出的分隔符

可以在 insert...后面使用 row format 来指定分隔符。示例如下：

```
hive (db01)> insert overwrite directory '/tb05'
           > row format delimited
           > fields terminated by '\t'
           > select id,name,year,month from tb05;
```

查看内容：

```
hive (db01)> dfs -cat /tb05/*;
101 Jack    2018    1
103 Mark    2018    1
102 Mary    2018    2
104 Rose    2018    2
```

可见数据已经根据\t 分隔开了。

5. 导出到不同的目录

可以使用如下语句：

```
From sometable
  Insert overwrite directory '...'
```

```
  Select * from ..
  Insert overwrite directory '...'
 ...
```

具体示例如下：

```
hive (db01)> from tb03
           > insert overwrite directory '/tb03-1'
           > select * where id>=500
           > insert overwrite directory '/tb03-2'
           > select * where id<500;
```

上面的语句完全可以根据分区来将所属分区对应的数据导出到不同的目录下。

另外，还有 Hadoop 命令导出和 Hive shell 命令导出等方式。

Hadoop 命令将 HDFS 中的数据导出到本地文件系统的指定目录中，比如：

```
hive(hivedwh)>dfs -get
/user/hive/warehouse/test/month=202109/000000_0
/opt/datas/export/test3.txt;
```

Hive shell 导出命令有两种方式：

（1）bin/hive -e HQL 语句 > filepath：将 HQL 语句的查询结果数据导出到指定目录下的文件中，示例如下：

```
[hadoop@server201 hive-3.1.3]$ bin/hive -e 'select * from test;' >
 /opt/datas/export/test4.txt;
```

（2）bin/hive -f 执行脚本 > filepath：将 HQL 语句存储在执行脚本文件中，将执行脚本文件的执行结果存储在指定目录下的文件中。例如，hivef.sql 脚本文件中存储 HQL 语句“select * from test;”，执行 hivef.sql 脚本文件中的查询语句，并将查询结果输出到 test5.txt 文件中：

```
[hadoop@server201 hive-3.1.3]$ bin/hive -f /hivef.sql >
 /opt/datas/export/test5.txt
```

第6章

Hive 查询

创建表并导入数据之后，即可对 Hive 表进行各种分析查询了。本章主要介绍 HQL 的各种查询方法，这是 Hive 数据仓库的重点内容之一。Hive 支持 SQL 数据定义语言（DML）中的几乎所有的功能，主要包括 Select 基本查询、Where 语句、分组语句、Join 语句、各种排序和抽样查询等。

6.1 Select...From 语句

普通的查询都比较简单，主要是对 array、map、struct 等数据类型的查询。

（1）准备数据，文本格式要完整：

```
2       Mary    blue,black,yellow    tel:1890099876,addr:china
1       Jack    red,green,blue       tel:1899089112,addr:sdjn
```

数据格式说明：

- 使用\t 进行分隔。
- 第一列为 id，第二列为名称，第三列为数组类型的爱好，第四列为 map 类型的联系方式。

（2）创建表：

```
create table tb06(
       id int,
       name string,
       hobbies array<String>,
       contact Map<String,String>
)
row format delimited
fields terminated by '\t'
collection items terminated by ','
```

```
map keys terminated by ':';
```

（3）加载数据：

```
hive (db01)> load data local inpath '/home/hadoop/tb06.txt' into table tb06;
Loading data to table db01.tb06
Table db01.tb06 stats: [numFiles=1, totalSize=51]
OK
Time taken: 0.515 seconds
```

（4）查看数据：

```
hive (db01)> select * from tb06;
OK
3   Mary    ["blue","black","yellow"]   {"tel":"1890099876","addr":"china"}
Time taken: 0.121 seconds, Fetched: 1 row(s)
```

已经看到，第三列和第四列已经显示了正确格式。

（5）查询 array 及 map：

```
    hive (db01)> select id,name,hobbies[0],hobbies[1],hobbies[2],
contact['tel'],contact['addr'] from tb06;
    OK
    3   Mary    blue    black   yellow  1890099876  china
    Time taken: 0.19 seconds, Fetched: 1 row(s)
```

6.2 Select 基本查询

Select 查询语句语法格式如下：

```
SELECT [ALL | DISTINCT] select_expr, select_expr, ...
FROM table_reference
[WHERE where_condition]
[GROUP BY col_list]
[HAVING having_condition]
[JOIN col_list ON …]
[ORDER BY col_list]
[SORT BY col_list]
[DISTRIBUTE BY col_list]
[CLUSTER BY col_list]
[TABLESAMPLE(BUCKET x OUT OF y ON col_list|RAND())]
[LIMIT number];
```

注意：

- HQL 语言对大小写不敏感。
- HQL 语句可以写在一行或者多行。
- 关键字不能被缩写也不能分行。
- 各子句一般要分行写。
- 使用缩进提高语句的可读性。

Hive 可以针对 Hive 表中的所有字段进行查询，也可以针对 Hive 表中特定的一个或几个字段进行查询，前者称为全表查询，后者称为特定列查询。

现有的数据二维表格格式如表 6-1 所示。

表6-1 数据二维表格

deptno	dname	buildingsno
100	数学系	2100
200	物理系	2200
300	化学系	2300
400	新闻系	2400
500	软件系	2500

首先，按照表 6-1 的数据生成一个 Hive 外部表 dept：

```
create external table dept (
deptno int,
dname string ,
buildingsno int
)ROW FORMAT DELIMITED FIELDS TERMINATED BY '\t' LINES TERMINATED BY '\n'
```

接下来进行查询。

（1）全表查询：

```
hive(hivedwh)>select * from dept;
```

（2）特定列查询：

```
hive(hivedwh)>select deptno, dname from dept;
```

为了便于分析计算，Hive 在查询时可以重命名字段，即重命名列的别名。列的别名紧跟列名，也可以在列名和别名之间加入关键字 AS。

例如，查询名称和部门：

```
hive(hivedwh)>select dname AS name, deptno dn from dept;
```

select 查询会返回符合条件的所有多行数据记录，limit 语句用于限制和设定返回的行数：

```
hive(hivedwh)>select * from dept limit 3;
```

这个典型的查询会返回多行数据，limit 子句限制返回 3 行记录。

6.3 Where 语句

Where 语句后面跟着一个条件表达式，该条件表达式可以是比较运算符表达式，也可以是逻辑运算符表达式，该条件表达式用于过滤数据。带有 Where 语句的查询返回一个有限的结果。Where 语句中不能使用字段别名。

Where 语句的作用是在对查询结果进行分组前，将不符合 Where 条件的记录去掉，即在分组之前过滤数据。使用 Where 条件显示特定的记录，条件中不能包含聚合函数。

以下为具体示例：

（1）创建一个文本文件 emp.txt，存储员工信息，表格内容如表 6-2 所示。

表6-2 员工信息表

empno	ename	gender	bday	area	score	deptno	scholarship
18999065	王述龙	男	1998-12-10	上海	98	100	2000
18007066	孙宇鹏	男	1999-11-17	沈阳	51	500	
18999141	王应龙	男	2000-02-04	沈阳	59	100	
18008158	张琼宇	女	1999-07-01	大连	89	200	
18999063	宋传涵	女	1999-07-20	上海	86	100	1000
18008009	李亚楠	女	1998-01-24	杭州	97	200	2000
18008026	侯楠楠	男	2000-01-29	北京	79	200	
18008027	陈姝元	女	1999-06-24	北京	96	200	1500
18009183	陆春宇	男	1998-01-18	沈阳	87	300	1000
18009173	孙云琳	女	1997-07-15	上海	56	300	
18008014	尤骞梓	女	1999-04-25	杭州	86	200	1000
18998002	张爱林	男	1999-05-16	北京	92	400	1500
18009019	曹雪东	男	2000-11-20	北京	78	300	
18998153	贾芸梅	女	2000-06-12	大连	88	400	1000
18007051	温勇元	男	1999-08-08	上海	65	500	
18998039	张微微	女	1998-01-27	北京	90	400	1500
18007063	李君年	男	1998-03-21	上海	78	500	
18007095	卢昱泽	女	1998-08-01	上海	57	500	
18007096	赵旭辉	男	1999-02-18	北京	75	500	
18009087	张矗年	男	1997-07-26	重庆	86	300	1000

（2）创建外部表 emp：

```
hive(hivedwh)>create external table if not exists emp(
empno int,
ename string,
gender string,
bday string,
area string,
score double,
deptno int,
scholarship double)
row format delimited fields terminated by '\t';
```

（3）向外部表 emp 中导入数据：

```
hive(hivedwh)>load data local inpath '/opt/datas/emp.txt' into table emp;
```

形成 Hive 表后，我们使用 Where 语句来做一下试验。

（1）查询成绩为 98 分的员工的所有信息：

```
hive(hivedwh)>select empno,ename,gender,bday,area,score,deptno
from emp where score =98;
```

（2）查询成绩为 90~100 分的员工信息：

```
hive(hivedwh)>select empno,ename,gender,area,score,deptno
from emp where score between 90 and 100;
```

（3）查询奖学金 scholarship 不为空的所有员工信息：

```
hive(hivedwh)>select empno,ename,gender,area,score,scholarship
from emp where scholarship is not null;
```

（4）查询成绩为 86 分或 96 分的员工信息：

```
hive(hivedwh)>select ename,gender,area,score,deptno
from emp where score IN (86, 96);
```

（5）查询成绩大于 90 分的所有员工信息：

```
hive(hivedwh)>select empno,ename,gender,bday,area,score,deptno from emp where
score >90;
```

6.4　Group By 语句

Group By 语句通常和聚合函数一起使用，按照一个或者多个字段进行分组，然后对每个组执行聚合操作。

1. 查询 emp 表每个部门的平均成绩

```
hive(hivedwh)>select deptno, avg(score) avg_score
from emp  group by deptno;

OK
deptno  avg_score
100     81.0
200     89.4
300     76.75
400     90.0
500     65.2
```

2. 查询 emp 表每个部门中男女性别的最好成绩

```
hive(hivedwh)>select deptno, gender, max(score) max_score
from emp group by deptno, gender;
OK
deptno   gender    max_score
100         女        86.0
200         女        97.0
300         女        56.0
400         女        90.0
500         女        57.0
100         男        98.0
200         男        79.0
300         男        87.0
400         男        92.0
500         男        78.0
```

Having 语句也用于限定返回的数据集。只有在 Group By 和 Having 语句中，才可以使用聚合函数。Having 语句在 Group By 语句之后，HQL 会在分组之后再计算 Having 语句，查询结果中只返回满足 Having 条件的结果。

（1）Where 语句针对表中的字段执行查询数据操作，而 Having 语句针对查询结果中的字段执行筛选数据操作。

（2）Where 语句后面不能使用聚合函数，而 Having 语句后面可以使用聚合函数。

（3）Having 语句只用于 Group By 分组统计语句。

例如，查询 emp 表中平均成绩大于 80 分的部门：

```
hive(hivedwh)>select deptno, avg(score) avg_score
from emp group by deptno
having avg_score > 80;
 OK
deptno      avg_score
100         81.0
200         89.4
400         90.0
```

6.5 Join 语句

1. 等值连接

Join 语句通过共同值组合来自两张表的特定字段，它是两张或更多的表组合的记录。Hive 支持通常的 SQL Join 语句，但是只支持等值连接，不支持非等值连接。

例如，根据 dept 表和 emp 表中的部门编号，查询员工编号、员工名称、部门编号和部门名称：

```
hive(hivedwh)>select e.empno, e.ename, d.deptno, d.dname
from emp e join dept d on e.deptno = d.deptno;
OK
e.empno      e.ename     d.deptno     d.dname
18999065     王述龙      100          数学系
18007066     孙宇鹏      500          软件系
18999141     王应龙      100          数学系
18008158     张琼宇      200          物理系
18999063     宋传涵      100          数学系
18008009     李亚楠      200          物理系
18008026     侯楠楠      200          物理系
18008027     陈姝元      200          物理系
18009183     陆春宇      300          化学系
18009173     孙云琳      300          化学系
18008014     尤骞梓      200          物理系
18998002     张爱林      400          新闻系
18009019     曹雪东      300          化学系
18998153     贾芸梅      400          新闻系
18007051     温勇元      500          软件系
18998039     张微微      400          新闻系
18007063     李君年      500          软件系
```

```
18007095    卢昱泽    500    软件系
18007096    赵旭辉    500    软件系
18009087    张矗年    300    化学系
```

使用表的别名不仅可以简化查询，还可以提高执行效率。

又如，合并 dept 表和 emp 表：

```
hive(hivedwh)>select e.empno, e.ename, d.deptno
from emp e join dept d on e.deptno = d.deptno;
```

2. 内连接

内连接只连接两个表中都存在的与连接条件相匹配的数据。内连接通过关键字 Inner Join 标识。例如：

```
hive(hivedwh)>select e.empno, e.ename, d.deptno
from emp e inner join dept d on e.deptno = d.deptno;
```

3. 外连接

左外连接是指 Join 操作符左边表中符合 Where 语句的所有记录将被返回。左外连接通过关键字 Left Outer Join 标识。例如：

```
hive(hivedwh)>select e.empno, e.ename, d.deptno
from emp e left outer join dept d on e.deptno = d.deptno;
```

右外连接是指 Join 操作符右边表中符合 Where 语句的所有记录将被返回。右外连接通过关键字 Right Outer Join 标识。例如：

```
hive(hivedwh)>select e.empno, e.ename, d.deptno
from emp e right outer join dept d on e.deptno = d.deptno;
```

注意：Where 语句在连接操作执行后才会执行，因此 Where 语句应只用于过滤那些非 Null 的列值，同时 On 语句中的分区过滤条件在外连接中是没用的，不过在内连接中是有效的。如果想在连接之前避免使用 On 语句条件和 Where 语句条件来过滤数据，那么可以使用嵌套查询。

满外连接将返回所有表中符合 Where 语句条件的所有记录。如果任一表的指定字段没有符合条件值，那么就使用 Null 值替代。满外连接通过关键字 Full Outer Join 标识。例如：

```
hive(hivedwh)>select e.empno, e.ename, d.deptno
from emp e full outer join dept d on e.deptno = d.deptno;
```

4. 左半连接

左半连接将返回左边表的记录，前提是其记录对于右边表满足 On 语句中的判断条件。对于常见的内连接来说，这是一种特殊的、优化了的情况。左半连接通过关键字 Left Semi Join 标识。例如：

```
hive(hivedwh)>select e.empno, e.ename, d.deptno
from emp e left semi join dept d on e.deptno = d.deptno;
```

一个查询可以连接两张以上的表。这里需要注意的是，连接 n 张表，至少需要 n-1 个连接条件。例如，连接 3 张表，至少需要 2 个连接条件。

5. 笛卡儿积 Join

笛卡儿积 Join 是一种连接，它把 Join 左边表的行数乘以右边表的行数的结果作为结果集，因此笛卡儿积 Join 会产生大量数据。和其他连接类型不同的是，笛卡儿积 Join 不是并行执行的，而且无法进行优化。

在 Hive 中，笛卡儿积 Join 在应用 Where 语句中的谓词条件前会先进行笛卡儿积计算，这个过程会消耗大量资源。如果设置属性 hive.mapred.mode 值为 strict，那么 Hive 会执行笛卡儿积 Join 查询，若无特别的要求，尽量不要使用笛卡儿积 Join。

笛卡儿积 Join 在一些情况下还是有用的，例如，有一张表为用户偏好，另一张表为新闻文章，同时有一个算法会推测出用户可能会喜欢读哪些文章，这个时候就需要笛卡儿积 Join 生成所有用户和所有网页的对应关系集合。

笛卡儿积 Join 产生的条件：

（1）省略连接条件。

（2）连接条件无效。

（3）所有表中的所有行互相连接。

以下是省略连接条件的情况，因此会采用笛卡尔积，具体实例操作为：

```
hive(hivedwh)>select empno, dname from emp, dept;
```

注意：连接谓词中不支持 or，以下查询是错误的。

```
hive(hivedwh)>select e.empno, e.ename, d.deptno
from emp e join dept d on e.deptno= d.deptno or e.ename=d.ename;
```

Hive 只支持等值连接、外连接和左半连接。Hive 不支持非相等的 Join 条件（通过其他方式实现，如 Left Outer Join），因为很难在 MapReduce 中实现这样的条件。Hive 可以 Join 两张以上的表。

6.6 排　　序

Hive 常用排序方法有 Order By、Sort By、Distribute By 和 Cluster By 等，下面将进行详细介绍。

6.6.1 Order By

Order By 按照一个或多个字段排序。

Hive 中的 Order By 和传统 SQL 中的 Order By 一样，对查询结果做全局排序，它会新启动一个任务进行排序，把所有数据放到同一个 Reduce 中进行处理。不管数据有多少，不管文件有多少，都启用一个 Reduce 进行处理。

数据量大的情况下将会消耗很长时间去执行排序，而且可能不会出结果，因此必须使用关键字 Limit 指定输出条数。

ASC（Ascend）表示升序（默认），DESC（Descend）表示降序。Order By 语句放在 Select 语句的结尾。

（1）查询学生信息按部门升序排列：

```
hive(hivedwh)>select ename,gender,bday,area,score,deptno
from emp order by deptno;

OK
ename    gender    bday       area    score   deptno
王述龙   男   1998-12-10      上海    98.0    100
宋传涵   女   1999-07-20      上海    86.0    100
王应龙   男   2000-02-04      沈阳    59.0    100
尤骞梓   女   1999-04-25      杭州    86.0    200
张琼宇   女   1999-07-01      大连    89.0    200
李亚楠   女   1998-01-24      杭州    97.0    200
侯楠楠   男   2000-01-29      北京    79.0    200
陈姝元   女   1999-06-24      北京    96.0    200
张矗年   男   1997-07-26      重庆    86.0    300
曹雪东   男   2000-11-20      北京    78.0    300
陆春宇   男   1998-01-18      沈阳    87.0    300
孙云琳   女   1997-07-15      上海    56.0    300
贾芸梅   女   2000-06-12      大连    88.0    400
张微微   女   1998-01-27      北京    90.0    400
张爱林   男   1999-05-16      北京    92.0    400
温勇元   男   1999-08-08      上海    65.0    500
李君年   男   1998-03-21      上海    78.0    500
卢昱泽   女   1998-08-01      上海    57.0    500
赵旭辉   男   1999-02-18      北京    75.0    500
孙宇鹏   男   1999-11-17      沈阳    51.0    500
```

（2）查询学生信息按成绩降序排列：

```
hive(hivedwh)>select ename,gender,bday,area,score,deptno
from emp order by score desc;
```

（3）字段别名排序。重命名一个字段，然后对重命名字段进行排序。例如，按照学生成绩的2倍排序：

```
hive(hivedwh)>select ename, score*2 twoscore
from emp order by twoscore;
```

（4）也可以对多字段同时进行排序。例如，按照部门和成绩两个字段升序排序：

```
hive(hivedwh)>select ename, deptno, score
from emp order by deptno, score;
ename    deptno score
王应龙   100    59.0
宋传涵   100    86.0
王述龙   100    98.0
侯楠楠   200    79.0
尤骞梓   200    86.0
张琼宇   200    89.0
陈姝元   200    96.0
李亚楠   200    97.0
孙云琳   300    56.0
曹雪东   300    78.0
张矗年   300    86.0
陆春宇   300    87.0
```

```
贾芸梅    400    88.0
张微微    400    90.0
张爱林    400    92.0
孙宇鹏    500    51.0
卢昱泽    500    57.0
温勇元    500    65.0
赵旭辉    500    75.0
李君年    500    78.0
```

6.6.2 Sort By

Sort By 是内部排序，会在每个 Reduce 中进行排序，单个 Reduce 出来的数据是有序的，但不保证全局有序。假设设置了 3 个 Reduce，那么这 3 个 Reduce 就会生成 3 个文件，每个文件都会按 Sort By 设置的条件排序，但是当这 3 个文件的数据合并在一起时，就不一定有序了。一般情况下，可以先进行 Sort By 内部排序，再进行全局排序，这样会提高排序效率。

使用 Sort By 可以先指定执行的 Reduce 个数（set mapreduce.job.reduces=<number>），对输出的数据再执行排序，即可以得到全部排序结果。

（1）设置 Reduce 个数为 3：

```
hive(hivedwh)>set mapreduce.job.reduces=3;
```

（2）查看设置的 Reduce 个数：

```
hive(hivedwh)>set mapreduce.job.reduces;
```

（3）根据部门编号降序查看学生信息：

```
hive(hivedwh)>select empno,ename,bday,area,score,deptno
from emp sort by deptno desc;
OK
empno       ename   bday         area      score      deptno
18998039    张微微  1998-01-27   北京      90.0       400
18009019    曹雪东  2000-11-20   北京      78.0       300
18999065    王述龙  1998-12-10   上海      98.0       100
18007051    温勇元  1999-08-08   上海      65.0       500
18007096    赵旭辉  1999-02-18   北京      75.0       500
18007095    卢昱泽  1998-08-01   上海      57.0       500
18009183    陆春宇  1998-01-18   沈阳      87.0       300
18009173    孙云琳  1997-07-15   上海      56.0       300
18008027    陈姝元  1999-06-24   北京      96.0       200
18008026    侯楠楠  2000-01-29   北京      79.0       200
18008009    李亚楠  1998-01-24   杭州      97.0       200
18999063    宋传涵  1999-07-20   上海      86.0       100
18007066    孙宇鹏  1999-11-17   沈阳      51.0       500
18007063    李君年  1998-03-21   上海      78.0       500
18998002    张爱林  1999-05-16   北京      92.0       400
18998153    贾芸梅  2000-06-12   大连      88.0       400
18009087    张矗年  1997-07-26   重庆      86.0       300
18008014    尤骞梓  1999-04-25   杭州      86.0       200
18008158    张琼宇  1999-07-01   大连      89.0       200
18999141    王应龙  2000-02-04   沈阳      59.0       100
```

（4）将查询结果导出到文件中（按照部门编号降序排序）：

```
hive(hivedwh)>insert overwrite local directory '/opt/datas/s_output'
select empno,ename,bday,area,score,deptno
from emp sort by deptno desc;
```

（5）查询导出文件中的数据。

目录 s_output 中生成了 3 个文件，分别为 000000_0、000001_0、000002_0。查看一下 000000_0 中的数据：

```
hadoop@server201:/opt/datas/s_output$ cat 000000_0

18007051 温勇元 1999-08-08 上海 65.0 500
18007096 赵旭辉 1999-02-18 北京 75.0 500
18007095 卢昱泽 1998-08-01 上海 57.0 500
18009183 陆春宇 1998-01-18 沈阳 87.0 300
18009173 孙云琳 1997-07-15 上海 56.0 300
18008027 陈姝元 1999-06-24 北京 96.0 200
18008026 侯楠楠 2000-01-29 北京 79.0 200
18008009 李亚楠 1998-01-24 杭州 97.0 200
18999063 宋传涵 1999-07-20 上海 86.0 100
```

6.6.3　Distribute By

Hive 中的 Distribute By 用于控制如何在 Map 端拆分数据给 Reduce 端：按照指定的字段把数据划分到不同的 Reduce 输出文件中，默认采用 Hash 算法。对于 Distribute By 分区排序，一定要多分配 Reduce 进行处理，否则无法看到 Distribute By 分区排序的效果。Hive 要求 Distribute By 语句写在 Sort By 语句之前。

Distribute By 和 Sort By 的使用场景主要包括：

- Map 输出的文件大小不均。
- Reduce 输出的文件大小不均。
- 小文件过多。
- 文件超大。

（1）先按照部门编号分区，再按照学生编号降序排序：

```
hive(hivedwh)>set mapreduce.job.reduces=3;
hive(hivedwh)>insert overwrite local directory '/opt/datas/d_output'
select empno,ename,bday,area,score,deptno
from emp distribute by deptno sort by empno desc;
```

（2）查询文件中的数据。

目录 10pt/datas/d_output 中生成了 3 个文件，分别为 000000_0、000001_0、000002_0，查看 000000_0 中的内容：

```
hadoop@server201:/opt/datas/d_output$ cat 000000_0

18009183 陆春宇 1998-01-18 沈阳 87.0 300
18009173 孙云琳 1997-07-15 上海 56.0 300
18009087 张矗年 1997-07-26 重庆 86.0 300
```

```
18009019 曹雪东 2000-11-20 北京 78.0 300
```

6.6.4 Cluster By

Cluster By 除了具有 Distribute By 的功能外，还兼具 Sort By 的功能。当 Distribute By 和 Sort By 字段相同时，可以使用 Cluster By 方式排序，但是只能是升序排序，不能降序排序。下面两种写法完全等价：

```
hive(hivedwh)>select empno,ename,bday,area,score,deptno
from emp cluster by deptno;
```

或：

```
hive(hivedwh)>select empno,ename,bday,area,score,deptno
from emp distribute  by deptno sort by deptno;
```

注意：按照部门编号分区，不一定就是固定的数值，可以是 200 号和 300 号部门被分到同一个分区里面。

6.7 抽样查询

当数据量特别大，对全部数据进行处理存在困难时，抽样查询就显得尤其重要了。抽样可以从被抽取的数据中估计和推断出整体的特性。Hive 支持桶表抽样查询、数据块抽样查询和随机抽样查询。

1. 桶表抽样查询

对于非常大的数据集，有时用户需要使用的是一个具有代表性的查询结果而不是全部结果，Hive 可以通过对桶表进行抽样查询来满足这个需求。

桶表抽样语法格式如下：

```
TABLESAMPLE(BUCKET x OUT OF y ON col_name | RAND())
```

TABLESAMPLE 语句允许用户编写用于数据抽样而不是整个表的查询，该语句出现在 From 语句中，可用于桶表中。桶表编号从 1 开始，col_name 表明抽取样本的字段，可以是非分区字段中的任意一个字段，或者使用随机函数 Rand()表明在整个行中抽取样本而不是单个字段。在 col_name 上桶表的行随机进入 1 到 y 个桶中，返回属于桶 x 的行。y 必须是桶表总桶数的倍数或者因子。Hive 根据 y 的大小决定抽样的比例。x 表示从哪个桶开始抽取，如果需要抽取多个桶，以后的桶号为当前桶号加上 y。需要注意的是，x 的值必须小于或等于 y 的值，否则会出现异常。

（1）查询桶表 test_b 中的全部数据：

```
hive(hivedwh)>select id,name from test_b;
OK
id    name
108   Ford
104   Linus
105   James
```

```
101   Bill
106   Steve
102   Dennis
107   Paul
103   Doug
```

（2）查询桶表 test_b 中的数据，抽取桶 1 数据：

```
hive(hivedwh)>select id,name
from test_b tablesample(bucket 1 out of 4 on id);

OK
Id  name
108  Ford
104  Linus
```

（3）查询桶表 test_b 中的数据，抽取桶 1 和桶 3 数据：

```
hive(hivedwh)>select id,name
from test_b tablesample(bucket 1 out of 2 on id);

OK
id  name
108  Ford
104  Linus
106  Steve
102  Dennis
```

（4）查询桶表 test_b 中的数据，随机抽取 4 个桶中的数据：

```
hive(hivedwh)>select id,name
from test_b tablesample(bucket 1 out of 4 on rand());

OK
id  name
106  Steve
102  Dennis
```

2. 数据块抽样查询

Hive 提供了一种按照数据块大小进行抽样查询的方式，这种方式可以按照输入路径下的数据块百分比、数据字节或者数据行数进行抽样。也就是说，该方式允许 Hive 按照数据总量的百分比、数据字节、数据行数三种方式之一随机抽取数据。

语法格式如下：

```
SELECT * FROM <Table_Name> TABLESAMPLE(N PERCENT|ByteLengthLiteral|N ROWS);
```

其中，ByteLengthLiteral 的取值为(Digit)+ (b | B | k | K| m | M| g | G)，表明数据的大小。

1）*按数据块百分比抽样*

按数据块百分比抽样允许抽取数据行数大小的至少 *n*%作为输入，支持 CombineHiveInput Format，而一些特殊的压缩格式是不能够被处理的，如果抽样失败，MapReduce 作业的输入将是整张表。因为是在 HDFS 块层级进行抽样，所以抽样粒度为块的大小，例如，如果块大小为 256MB，那么即使输入的 *n*%仅为 100MB，也会得到 256MB 的数据。例如：

```
hive(hivedwh)>select ename, bday, score
from emp tablesample(10 percent);

ename    bday        score
王述龙    1998-12-10 98.0
孙宇鹏    1999-11-17 51.0
```

2）按数据大小抽样

按数据大小抽样方式的最小抽样单元是一个 HDFS 数据块。如果数据大小小于普通的块大小（128MB），那么会返回所有的行。例如：

```
hive(hivedwh)>select ename, bday, score
from emp tablesample(1b);
OK
ename   bday        score
王述龙   1998-12-10  98.0

hive(hivedwh)>select ename, bday, score
from emp tablesample(1k);
OK
ename    bday        score
王述龙    1998-12-10 98.0
孙宇鹏    1999-11-17 51.0
王应龙    2000-02-04 59.0
```

3. 随机抽样查询

使用 Rand()函数可以进行随机抽样查询，Limit 关键字限制抽样返回的数据条数，其中 Rand()函数前的 Distribute By 和 Sort By 关键字可以保证数据在 Map 和 Reduce 阶段是随机分布的。

例如，随机抽取 emp 表中的 10 条数据：

```
hive(hivedwh)>select empno,ename,bday
from emp distribute by rand() sort by rand() limit 10;
OK
empno       ename   bday
18008158    张琼宇   1999-07-01
18008009    李亚楠   1998-01-24
18008027    陈姝元   1999-06-24
18009183    陆春宇   1998-01-18
18008026    侯楠楠   2000-01-29
18998039    张微微   1998-01-27
18007066    孙宇鹏   1999-11-17
18009019    曹雪东   2000-11-20
18009173    孙云琳   1997-07-15
18998153    贾芸梅   2000-06-12
```

第 7 章

Hive 函数

Hive 函数分为两类，分别是内置函数和自定义函数。

7.1 查看系统内置函数

（1）查看系统自带的函数：

```
hive> show functions;
```

（2）显示自带的函数的用法：

```
hive> desc function upper;
```

（3）详细显示自带的函数的用法：

```
hive> desc function extended upper;
```

7.2 常用内置函数

Hive 内部支持大量的内置函数，可以通过 SHOW FUNCTIONS 命令查看这些内置函数。灵活地运用 Hive 提供的函数能够极大地节省数据分析成本。Hive 函数可以分为数学函数、集合函数、类型转换函数等多种类型，下面介绍 Hive 中常用的几个内置函数。

1. NVL 函数（空字段赋值）

NVL 函数给值为 NULL 的数据赋值，它的格式是 NVL(value, default_value)。它的功能是如果 value 为 NULL，则 NVL 函数返回 default_value 的值，否则返回 value 的值；如果两个参数都为 NULL，则返回 NULL。

2. CASE WHEN 判断

用法：CASE 字段 WHEN 判断条件 THEN 进行的操作 ELSE 不满足条件进行的操作 END。CASE WHEN 对某一字段进行按条件操作，类似于 Java 中的 if…else…语句。

3. 行转列（拼接函数）

拼接函数有以下 4 种。

（1）CONCAT(string A/col, string B/col…)：返回输入字符串连接后的结果，支持任意个输入字符串拼接。

（2）CONCAT_WS(separator, str1, str2,…)：它是一种特殊形式的 CONCAT()。第一个参数是其余参数间的分隔符。分隔符可以是与其余参数一样的字符串。如果分隔符是 NULL，那么返回值也将为 NULL。这个函数会跳过分隔符参数后的任何 NULL 和空字符串。分隔符将被添加到被连接的字符串之间。

注意：CONCAT_WS 的参数必须是 string 或 array。

（3）COLLECT_SET(col)：此函数只接收基本数据类型，它的主要作用是将某字段的值进行去重汇总，产生 array 类型字段。

（4）COLLECT_LIST(col)：此函数只接收基本数据类型，它的主要作用是将某字段的值进行不去重汇总，产生 array 类型字段。

示例如下：

```
SELECT t1.c_b , CONCAT_WS("|",collect_set(t1.name))
FROM (
    SELECT NAME ,CONCAT_WS(',',constellation,blood_type) c_b
    FROM person_info
)t1
GROUP BY t1.c_b
```

4. 列转行

列转行函数说明如下：

- SPLIT(str, separator): 将字符串按照后面的分隔符 separator 进行切割，转换成字符 array。
- EXPLODE(col): 将 Hive 的一列中复杂的 array 或者 map 结构拆分成多行。
- LATERAL VIEW udtf(expression) tableAlias AS columnAlias: LATERAL VIEW 用于和 SPLIT、EXPLODE 等 UDTF 一起使用，它能够将一行数据拆分成多行数据，在此基础上可以对拆分后的数据进行聚合。LATERAL VIEW 首先为原始表的每行调用 UDTF, UTDF 会把一行拆分成一行或者多行，LATERAL VIEW 再把结果进行组合，产生一张支持别名表的虚拟表。

LATERAL VIEW 用于维持查询表和拆分后的虚拟表的联系。

LATERAL VIEW 查询数据示例如下：

```
SELECT movie,category_name
FROM movie_info
```

```
LATERAL VIEW
explode(split(category,",")) movie_info_tmp  AS category_name ;
```

5. 窗口函数（开窗函数）

窗口函数说明如下：

- OVER(): 指定分析函数工作的数据窗口大小，这个数据窗口大小可能会随着行的变化而变化。
- ROWS BETWEEN 行开始位置 AND 行结束位置。
- CURRENT ROW: 当前行。
- n PRECEDING: 往前 *n* 行数据。
- n FOLLOWING: 往后 *n* 行数据。
- UNBOUNDED: 无边界。
- UNBOUNDED PRECEDING: 前无边界，表示从前面的起点。
- UNBOUNDED FOLLOWING: 后无边界，表示到后面的终点。
- LAG(col,n,default_val): 往前第 *n* 行数据。
- LEAD(col,n, default_val): 往后第 *n* 行数据。
- FIRST_VALUE (col,true/false): 当前窗口下的第一个值，若第二个参数为 true，则表示跳过空值。
- LAST_VALUE (col,true/false): 当前窗口下的最后一个值，若第二个参数为 true，则表示跳过空值。
- NTILE(n): 把有序窗口的行分发到指定数据的组中，各个组有编号，编号从 1 开始，对于每一行，NTILE 返回此行所属的组的编号。注意：*n* 必须为 int 类型。

（1）查询在 2017 年 4 月份下过订单的顾客及总人数：

```
select name,count(*) over ()
from business
where substring(orderdate,1,7) = '2017-04'
group by name;
```

（2）查询顾客的购买明细及月购买总额：

```
select
    name,
    orderdate,
    cost,
    sum(cost) over(partition by name,month(orderdate)) name_month_cost
from business;
```

（3）将每个顾客的 cost 按照日期进行累加：

```
select name,orderdate,cost,
    sum(cost) over() as sample1,--所有行相加
    sum(cost) over(partition by name) as sample2,--按 name 分组，组内数据相加
    sum(cost) over(partition by name order by orderdate) as sample3,--按 name
分组，组内数据累加
    sum(cost) over(partition by name order by orderdate rows between UNBOUNDED
PRECEDING and current row ) as sample4 ,--和 sample3 一样，由起点到当前行的聚合
```

```
    sum(cost) over(partition by name order by orderdate rows between 1 PRECEDING
and current row) as sample5, --当前行和前面一行做聚合
    sum(cost) over(partition by name order by orderdate rows between 1 PRECEDING
AND 1 FOLLOWING ) as sample6,--当前行和前面一行及后面一行
    sum(cost) over(partition by name order by orderdate rows between current row
and UNBOUNDED FOLLOWING ) as sample7 --当前行及后面所有行
from business;
```

rows 必须跟在 order by 子句之后，使用固定的行数来限制分区中的数据行数量。

（4）查询顾客购买明细以及上次的购买时间和下次购买时间：

```
select
    name,orderdate,cost,
    lag(orderdate,1,'1970-01-01') over(PARTITION by name order by orderdate)
prev_time,
    lead(orderdate,1,'1970-01-01') over(PARTITION by name order by orderdate)
next_time
from business;
```

（5）查询顾客每个月第一次的购买时间和每个月的最后一次的购买时间：

```
select
    name,
    orderdate,
    cost,
    FIRST_VALUE(orderdate) over(partition by name,month(orderdate) order by
orderdate rows between UNBOUNDED PRECEDING and UNBOUNDED FOLLOWING) first_time,
    LAST_VALUE(orderdate) over(partition by name,month(orderdate) order by
orderdate rows between UNBOUNDED PRECEDING and UNBOUNDED FOLLOWING) last_time
from business;
```

（6）查询前 20%时间的订单信息：

```
select * from (
    select name,orderdate,cost, ntile(5) over(order by orderdate) sorted
    from business
) t
where sorted = 1;
```

6. 排序函数

排序函数说明如下：

- RANK(): 排序相同时会重复（考虑并列、会跳号），总数不会变。
- DENSE_RANK(): 排序相同时会重复（考虑并列、不跳号），总数会减少。
- ROW_NUMBER(): 会根据顺序计算（不考虑并列、不跳号）。

排序函数示例如下：

```
select name,
    subject,
    score,
    rank() over(partition by subject order by score desc) rp,
    dense_rank() over(partition by subject order by score desc) drp,
```

```
    row_number() over(partition by subject order by score desc) rmp
from score;
```

7.3 Hive的其他函数

本节将介绍Hive的除7.2节的内置函数之外的其他常用函数及其实例。

7.3.1 准备数据

（1）创建一张表，并指定每一行的分隔符：

```
hive> create table stud(id string,name string,sex string,age int)
    > row format delimited fields terminated by '\t';
OK
Time taken: 0.233 seconds
```

（2）在Linux上创建一个文本文件用于保存以下数据：

```
[hadoop@server201 ~]$ vim stud.txt
```

创建完成文件以后，将它上传到HDFS并查看里面的内容：

```
hdfs dfs -cat /test/stud.txt
T001    Jack    1   23
T002    Mary    0   24
T003    Rose    1   34
T004    Alex    1   23
T100    Smith   1   29
T005    Jim     0   34
T000    张三    1   38
```

（3）使用load data inpath将HDFS中的文件导入刚创建的表中：

```
load data inpath '/test/stud.txt' into table stud;
```

导入以后的数据将会被保存到Hive的目录下。

使用load data加载完成以后，源数据将被删除，可以再次使用load data加载新的数据。load data的语法如下：

```
LOAD DATA [LOCAL] INPATH 'filepath' [OVERWRITE] INTO TABLE tablename [PARTITION
(partcol1=val1, partcol2=val2 ...)]
```

7.3.2 其他函数的使用

Hive中内置函数众多，本节将讲解7.2节给出的函数之外的其他函数。我们主要通过示例来简单演示它们的用法，方便读者使用时查阅。

1）show functions

查看所有的内置函数以及用户加载进来的自定义函数。例如：

```
hive> show functions;
OK
```

```
acos
hive> describe function acos;
OK
acos(x) - returns the arc cosine of x if -1<=x<=1 or NULL otherwise
Time taken: 0.051 seconds, Fetched: 1 row(s)
```

2）add_months

给某个日期添加几个月。例如：

```
hive> describe function add_months;
OK
add_months(start_date, num_months) - Returns the date that is num_months after start_date.
Time taken: 0.041 seconds, Fetched: 1 row(s)
hive> select add_months('2009-09-09',2);
OK
2009-11-09
Time taken: 0.164 seconds, Fetched: 1 row(s)
```

3）array

创建一个数组。例如：

```
hive> describe function array;
OK
array(n0, n1...) - Creates an array with the given elements
Time taken: 0.027 seconds, Fetched: 1 row(s)
hive> select array(1,2,3,4);
OK
[1,2,3,4]
Time taken: 0.186 seconds, Fetched: 1 row(s)
hive> select array('Jack','Mary','Alex');
OK
["Jack","Mary","Alex"]
Time taken: 0.159 seconds, Fetched: 1 row(s)
```

4）array_contains

判断一个数组中是否存在另一个元素。例如：

```
hive> describe function array_contains;
OK
array_contains(array, value) - Returns TRUE if the array contains value.
Time taken: 0.029 seconds, Fetched: 1 row(s)
hive> select array_contains(array(1,2,3),1);
OK
true
Time taken: 0.164 seconds, Fetched: 1 row(s)
hive> select array_contains(array('Jack','Mary'),'Jack');
OK
true
Time taken: 0.181 seconds, Fetched: 1 row(s)
```

5）ascii

将字符串转成 ascii 码。例如：

```
hive> select ascii('a');
```

```
OK
97
Time taken: 0.132 seconds, Fetched: 1 row(s)
hive> select ascii('abc');
OK
97
Time taken: 0.13 seconds, Fetched: 1 row(s)
```

6）avg

求平均值。之前已经说过，执行Hive的SQL就是执行一个MapReduce程序。下面示例对7.3.1节准备的数据执行avg操作。

（1）查询所有人的平均成绩，这个avg运算将开始一个MapReduce程序：

```
select  avg(score) avg_score from emp
```

（2）在avg中使用分组group by：

```
select deptno, avg(score) avg_score from emp  group by deptno
```

7）base64

将二进制格式的数据转换成base64编码的字符串。例如：

```
select base64(cast('abcd' as binary))
```

8）between

between跟and搭配，表示列在两个数值之间，而且头尾两个数值也在范围之内。例如：

```
select * from emp where score between 80 and 90
```

9）bin

将一个bigint类型转成二进制类型。例如：

```
hive> select bin(90);
OK
1011010
Time taken: 0.145 seconds, Fetched: 1 row(s)
```

10）case

条件判断函数，与When语句联用，用法为Case a When b Then c [When d Then e]* [Else f] End。可以这样理解：如果a等于b，那么返回c；如果a等于d，那么返回e；否则返回f。

以下例子为查询部门员工统计信息，主要查询男女员工的人数：

```
select deptno,
  sum(case gender when '男' then 1 else 0 end) male_c,
  sum(case gender when '女' then 1 else 0 end) female_c
from emp
group by deptno;
```

11）ceil

本函数接收small int，返回大于当前数的最小整数。例如：

```
hive> describe function ceil;
OK
ceil(x) - Find the smallest integer not smaller than x
```

```
Time taken: 0.021 seconds, Fetched: 1 row(s)
hive> select ceil(3.1);
OK
4
Time taken: 0.14 seconds, Fetched: 1 row(s)
hive> select ceil(-3.1);
OK
-3
Time taken: 0.123 seconds, Fetched: 1 row(s)
```

12）ceiling

本函数是返回不小于当前数的最小值，经过测试与上面 ceil 函数的结果相同。例如：

```
hive> describe function ceiling;
OK
ceiling(x) - Find the smallest integer not smaller than x
Time taken: 0.038 seconds, Fetched: 1 row(s)
hive> select ceiling(3.9);
OK
4
Time taken: 0.123 seconds, Fetched: 1 row(s)
hive> select ceiling(4.0);
OK
4
Time taken: 0.123 seconds, Fetched: 1 row(s)
hive> select ceiling(0.3);
OK
1
Time taken: 0.141 seconds, Fetched: 1 row(s)
```

13）collect_list

返回拥有重复数据的某个列的一个数组。用法如下：

```
hive> describe function collect_list;
OK
collect_list(x) - Returns a list of objects with duplicates
Time taken: 0.046 seconds, Fetched: 1 row(s)
```

对于表，执行 collect_list 之后返回 list 列表：

```
hive> select collect_list(age) from stud;
[23,24,34,23,29,34,38,90,34,34]
```

14）collect_set

返回不重复的集合。列转行专用函数，有时为了字段拼接效果，多和 concat_ws()函数联用。以下为其简单用法：

```
hive> describe function collect_set;
OK
collect_set(x) - Returns a set of objects with duplicate elements eliminated
Time taken: 0.048 seconds, Fetched: 1 row(s)
```

示例：

```
hive> select collect_set(age) from stud;
[23,24,34,29,38,90]
```

15）compute_stats

返回一个列的统计信息。比如对于 age 列执行查询：

```
Hive> select compute_stats(age,1) from stud;
{"columntype":"Long","min":23,"max":90,"countnulls":1,"numdistinctvalues":41}
```

参数说明：

- columntype：类型。
- min：最小值。
- max：最大值。
- countnulls：空值。
- numdistinctvalues：非重复值的数目。

16）concat

字符串连接。用法如下：

```
hive> describe function concat;
OK
concat(str1, str2, ... strN) - returns the concatenation of str1, str2, ... strN
or concat(bin1, bin2, ... binN) - returns the concatenation of bytes in binary data
bin1, bin2, ... binN
Time taken: 0.05 seconds, Fetched: 1 row(s)
```

示例：

```
hive> select concat('Jack','Mary');
OK
JackMary
Time taken: 0.084 seconds, Fetched: 1 row(s)
```

17）concat_ws

将数组进行串联。用法如下：

```
hive> describe function concat_ws;
OK
concat_ws(separator, [string | array(string)]+) - returns the concatenation of
the strings separated by the separator.
Time taken: 0.017 seconds, Fetched: 1 row(s)
```

示例：

```
hive> select concat_ws('#','Jack','Mary','alex');
OK
Jack#Mary#alex
Time taken: 0.1 seconds, Fetched: 1 row(s)
```

以下为将数据表中的 name 列转成一行显示：

```
hive> select concat_ws(',',collect_set(name)) from stud;
Jack,Mary,Rose,Alex,Smith,Jim,张三,Tom
```

18）conv

用于将一个数字转成指定的进制。例如：

```
hive> describe function conv;
OK
conv(num, from_base, to_base) - convert num from from_base to to_base
Time taken: 0.019 seconds, Fetched: 1 row(s)
hive> select conv(15,10,16);
OK
F
Time taken: 0.092 seconds, Fetched: 1 row(s)
```

19）corr 返回皮尔森相关系数

返回分组组内一对元素的皮尔森相关系数。返回类型为 double 类型。用法如下：

```
corr(col1, col2)
```

示例：

```
select id,corr(col1, col2) from t group by id
```

20）count

- count(*): 返回行的个数，包括 NULL 值的行。
- count(Expr): 返回指定字段的非空值行的个数。
- count(Distinct Expr[, Expr_.]): 返回指定字段的不同的非空值行的个数。

例如：

```
hive> select count(*) from stud;
OK
7
```

21）current_database

返回当前的数据库名称。例如：

```
hive> select current_database();
OK
default
Time taken: 0.074 seconds, Fetched: 1 row(s)
```

22）current_date

返回当前时间。例如：

```
hive> select current_date();
OK
2018-07-03
Time taken: 0.083 seconds, Fetched: 1 row(s)
```

23）current_timestamp

返回当前时间戳。例如：

```
hive> select current_timestamp();
OK
2018-07-03 15:46:27.491
Time taken: 0.088 seconds, Fetched: 1 row(s)
```

24）current_user

返回 Linux 当前用户。例如：

```
hive> select current_user();
OK
hadoop
Time taken: 0.086 seconds, Fetched: 1 row(s)
```

25）date_add

两个日期相加。例如：

```
hive> describe function date_add;
OK
date_add(start_date, num_days) - Returns the date that is num_days after
start_date.
Time taken: 0.023 seconds, Fetched: 1 row(s)
hive> select date_add('2009-09-10',39);
OK
2009-10-19
Time taken: 0.096 seconds, Fetched: 1 row(s)
```

26）date_format

时间格式化。例如：

```
hive> describe function date_format;
OK
date_format(date/timestamp/string, fmt) - converts a date/timestamp/string to
a value of string in the format specified by the date format fmt.
Time taken: 0.023 seconds, Fetched: 1 row(s)
hive> select date_format(current_timestamp,'yyyy/MM');
OK
2018/07
Time taken: 0.091 seconds, Fetched: 1 row(s)
```

27）datediff

返回两个日期相差的天数。例如：

```
hive> select datediff('2009-09-12','2018-09-23');
OK
-3298
Time taken: 0.111 seconds, Fetched: 1 row(s)
hive> select datediff('2018-09-12','2018-09-23');
OK
-11
Time taken: 0.105 seconds, Fetched: 1 row(s)
```

28）day

返回日期。例如：

```
hive> describe function day;
OK
day(param) - Returns the day of the month of date/timestamp, or day component
of interval
Time taken: 0.024 seconds, Fetched: 1 row(s)
hive> select day('2009-09-23');
OK
23
Time taken: 0.098 seconds, Fetched: 1 row(s)
```

29）dayofmonth

获取一天是一个月中的第几天。用法如下：

```
dayofmonth(date)
```

30）decode

比较表达式和搜索字是否匹配，如果匹配，则返回结果；如果不匹配，则返回 default 值；如果未定义 default 值，则返回空值。例如：

```
decode(status, '01', 1, '02', 2, 0 )
```

如果字段 status 的值为 01，则返回 1；如果字段 status 的值为 02，则返回 2；否则返回 0。

31）div

两数相除，只接收 bigint 类型。例如：

```
hive> select 10 div 2;
OK
5
Time taken: 0.109 seconds, Fetched: 1 row(s)
```

32）exp

e 的 *n* 次方。用法如下：

```
hive> desc function exp;
OK
exp(x) - Returns e to the power of x
Time taken: 0.01 seconds, Fetched: 1 row(s)
```

33）explode

将一个数组解析成多行。用法如下：

```
hive> desc function explode;
OK
explode(a) - separates the elements of array a into multiple rows, or the elements
of a map into multiple rows and columns
Time taken: 0.013 seconds, Fetched: 1 row(s)
```

示例如下：

```
hive> select explode(array(1,2,3));
OK
1
2
3
Time taken: 0.112 seconds, Fetched: 3 row(s)
hive> select explode(array('Jack','Mary'));
OK
Jack
Mary
Time taken: 0.092 seconds, Fetched: 2 row(s)
```

34）factorial

返回一个数（0~20）的阶乘。例如：

```
hive> desc function factorial;
OK
factorial(int) - Returns n factorial. Valid n is [0..20].
Time taken: 0.045 seconds, Fetched: 1 row(s)
hive> select factorial(10);
OK
3628800
Time taken: 0.136 seconds, Fetched: 1 row(s)
```

35）format_number

将数字格式化。例如：

```
hive> desc function format_number;
OK
format_number(X, D) - Formats the number X to a format like '#,###,###.##', rounded to D decimal places, and returns the result as a string. If D is 0, the result has no decimal point or fractional part. This is supposed to function like MySQL's FORMAT
Time taken: 0.03 seconds, Fetched: 1 row(s)
hive> select format_number(3.45678,2);
OK
3.46
Time taken: 0.161 seconds, Fetched: 1 row(s)
```

36）greatest

求最大值。例如：

```
hive> desc function greatest;
OK
greatest(v1, v2, ...) - Returns the greatest value in a list of values
Time taken: 0.005 seconds, Fetched: 1 row(s)
hive> select greatest(3,9,10,100);
OK
100
Time taken: 0.141 seconds, Fetched: 1 row(s)
```

37）if

条件判断。用法为 if(条件,条件为 true 时返回此值,条件为 false 时返回此值)。例如：

```
hive> desc function if;
OK
IF(expr1,expr2,expr3) - If expr1 is TRUE (expr1 <> 0 and expr1 <> NULL) then IF() returns expr2; otherwise it returns expr3. IF() returns a numeric or string value, depending on the context in which it is used.
Time taken: 0.038 seconds, Fetched: 1 row(s)
hive> select if(1=1,111,222);
OK
111
Time taken: 0.175 seconds, Fetched: 1 row(s)
hive> select if(1=2,111,222);
OK
222
Time taken: 0.111 seconds, Fetched: 1 row(s)
```

38）Index

从数组中查找指定下标的元素。例如：

```
hive> desc function index;
OK
index(a, n) - Returns the n-th element of a
Time taken: 0.021 seconds, Fetched: 1 row(s)
hive> select index(array('Jack','Mary'),1);
OK
Mary
Time taken: 0.131 seconds, Fetched: 1 row(s)
```

39）initcap

按空格分开的首字母大写。例如：

```
hive> desc function initcap;
OK
initcap(str) - Returns str, with the first letter of each word in uppercase,
all other letters in lowercase. Words are delimited by white space.
Time taken: 0.017 seconds, Fetched: 1 row(s)
hive> select initcap('Jack mary is friends');
OK
Jack Mary Is Friends
Time taken: 0.156 seconds, Fetched: 1 row(s)
Inline
hive> desc function inline;
OK
inline( ARRAY( STRUCT()[,STRUCT()] - explodes and array and struct into a table
Time taken: 0.015 seconds, Fetched: 1 row(s)
```

40）Instr

返回第一个子字符串的索引。例如：

```
hive> desc function instr;
OK
instr(str, substr) - Returns the index of the first occurance of substr in str
Time taken: 0.033 seconds, Fetched: 1 row(s)
hive> select instr('Jack And Mary','Jack');
OK
1
Time taken: 0.113 seconds, Fetched: 1 row(s)
hive> select instr('Jack And Mary','And');
OK
6
Time taken: 0.167 seconds, Fetched: 1 row(s)
```

41）map

该函数定义一个 map 对象。例如：

```
hive> select map('name','jack');
OK
{"name":"jack"}
Time taken: 0.164 seconds, Fetched: 1 row(s)
hive> select map('name','jack','age',34);
```

```
OK
{"name":"jack","age":"34"}
Time taken: 0.117 seconds, Fetched: 1 row(s)
hive> desc function map;
OK
map(key0, value0, key1, value1...) - Creates a map with the given key/value pairs
Time taken: 0.024 seconds, Fetched: 1 row(s)
```

42）struct

将多个字段构建为struct结构。例如：

```
hive> desc function struct;
OK
struct(col1, col2, col3, ...) - Creates a struct with the given field values
Time taken: 0.032 seconds, Fetched: 1 row(s)
hive> select struct('Jack','Mary');
OK
{"col1":"Jack","col2":"Mary"}
Time taken: 0.122 seconds, Fetched: 1 row(s)
hive> select struct(name) from stud;
OK
{"col1":"Jack"}
{"col1":"Mary"}
{"col1":"Rose"}
{"col1":"Alex"}
{"col1":"Smith"}
{"col1":"Jim"}
{"col1":"张三"}
{"col1":"Jack"}
{"col1":"Tom"}
{"col1":"Mary"}
{"col1":"Rose"}
Time taken: 0.108 seconds, Fetched: 11 row(s)
hive> select struct(name,age) from stud;
OK
{"col1":"Jack","col2":23}
{"col1":"Mary","col2":24}
{"col1":"Rose","col2":34}
{"col1":"Alex","col2":23}
{"col1":"Smith","col2":29}
{"col1":"Jim","col2":34}
{"col1":"张三","col2":38}
{"col1":"Jack","col2":90}
{"col1":"Tom","col2":34}
{"col1":"Mary","col2":null}
{"col1":"Rose","col2":34}
Time taken: 0.106 seconds, Fetched: 11 row(s)
```

7.3.3　显示某个函数的帮助信息

使用“describe function <函数名称>”可以显示某个函数的帮助信息。例如：

```
hive> describe function upper;
OK
upper(str) - Returns str with all characters changed to uppercase
```

```
Time taken: 0.045 seconds, Fetched: 1 row(s)
```

7.4 自定义函数

当 Hive 提供的内置函数无法满足我们的业务处理需要时，就可以考虑使用用户自定义函数（UDF，User-Defined Function）。

用户自定义函数分为以下 3 种：

（1）UDF（User-Defined-Function）：输入输出值为一进一出。

（2）UDAF（User-Defined Aggregation Function）：用户自定义聚合函数，输入输出值为多进一出。类似于 count、max、min 等函数。

（3）UDTF（User-Defined Table-Generating Functions）：用户自定义表生成函数，输入输出值为一进多出。比如 lateral view explode()。

用户自定义函数的官方文档地址为 https://cwiki.apache.org/confluence/display/Hive/HivePlugins。

自定义函数编程步骤如下：

步骤01 继承 Hive 提供的类：

```
org.apache.hadoop.hive.ql.udf.generic.GenericUDF
org.apache.hadoop.hive.ql.udf.generic.GenericUDTF;
```

步骤02 实现类中的抽象方法。

步骤03 在 Hive 的命令行窗口创建函数。

添加 JAR：

```
add jar linux_jar_path
```

创建 function：

```
create [temporary] function [dbname.]function_name AS class_name;
```

步骤04 在 Hive 的命令行窗口删除函数：

```
drop [temporary] function [if exists] [dbname.]function_name;
```

7.4.1 Hive 自定义 UDF 的过程

下面我们来详细介绍一下 Hive 自定义 UDF 的过程。

1. 编写 UDF

要求自定义一个 UDF 实现计算给定字符串的长度，例如：

```
hive(default)> select my_len("abcd");
```

返回值为 4。

（1）使用 IDEA 创建一个 Maven 工程，命名为 myudf。

（2）导入依赖：

```
<dependencies>
    <dependency>
        <groupId>org.apache.hive</groupId>
        <artifactId>hive-exec</artifactId>
        <version>3.1.2</version>
    </dependency>
</dependencies>
```

（3）创建一个类 MyLength：

```
package com.hadoop.hive.udf;
import org.apache.hadoop.hive.ql.exec.UDFArgumentException;
import org.apache.hadoop.hive.ql.exec.UDFArgumentLengthException;
import org.apache.hadoop.hive.ql.exec.UDFArgumentTypeException;
import org.apache.hadoop.hive.ql.metadata.HiveException;
import org.apache.hadoop.hive.ql.udf.generic.GenericUDF;
import org.apache.hadoop.hive.serde2.objectinspector.ObjectInspector;
import org.apache.hadoop.hive.serde2.objectinspector.primitive.
PrimitiveObjectInspectorFactory;
/**
 * 自定义 UDF，需要继承 GenericUDF 类
 * 需求：计算指定字符串的长度
 */
public class MyLength extends GenericUDF {
    /**
     * 初始化方法，里面要做 3 件事
     * 1.约束函数传入参数的个数
     * 2.约束函数传入参数的类型
     * 3.约束函数返回值的类型
     * @param arguments  函数传入参数的类型
     * @return
     * @throws UDFArgumentException
     */
    @Override
    public ObjectInspector initialize(ObjectInspector[] arguments) throws
UDFArgumentException {
        //1.约束函数传入参数的个数
        if (arguments.length != 1) {
            throw new UDFArgumentLengthException("Input Args Num Error,You can
only input one arg...");
        }
        //2.约束函数传入参数的类型
        if
(!arguments[0].getCategory().equals(ObjectInspector.Category.PRIMITIVE)) {
            throw new UDFArgumentTypeException(0,"Input Args Type Error,You can
only input PRIMITIVE Type...");
        }
        //3.约束函数返回值的类型
        return PrimitiveObjectInspectorFactory.javaIntObjectInspector;
    }
    /**
     * 函数逻辑处理方法
```

```
     * @param arguments  函数传入参数的值
     * @return
     * @throws HiveException
     */
    @Override
    public Object evaluate(DeferredObject[] arguments) throws HiveException {
        //获取函数传入参数的值
        Object o = arguments[0].get();
        //将 object 转换成字符串
        int length = o.toString().length();
        //因为在上面的初始化方法里面已经对函数返回值类型做了约束，必须返回一个 int 类型
        //所以我们要在这个地方直接返回 length
        return length;
    }
    /**
     * 返回显示字符串方法，这个方法不用管，直接返回一个空字符串
     * @param children
     * @return
     */
    @Override
    public String getDisplayString(String[] children) {
        return "";
    }
}
```

2. 创建临时函数

（1）将 myudf 项目打成 JAR 包上传到服务器/opt/module/hive/datas/myudf.jar。

（2）将 JAR 包添加到 Hive 的 classpath，临时生效：

```
hive (default)> add jar /opt/module/hive/datas/myudf.jar;
```

（3）创建临时函数并与开发好的 Java class 关联：

```
hive (default)> create temporary function my_len as
"com.hadoop.hive.udf.MyLength";
```

（4）此时即可在 HQL 中使用自定义的临时函数：

```
hive (default)> select ename,my_len(ename) ename_len from emp;
```

（5）删除临时函数：

```
hive (default)> drop  temporary function my_len;
```

注意：临时函数只跟会话有关系，跟库没有关系。只要创建临时函数的会话不断，则在当前会话下，任意一个库都可以使用这个临时函数，其他会话全都不能使用。

3. 创建永久函数

（1）在 Hive 主目录下面创建 auxlib 目录：

```
[hadoop@hadoop102 hive]$ mkdir auxlib
```

（2）将 JAR 包上传到$HIVE_HOME/auxlib 下，然后重启 Hive。

（3）创建永久函数：

```
hive (default)> create function 自定义函数名 as "udf方法的全类名";
```

例如，创建一个名称为 my len2 的函数：

```
create function my len2 as "com.hadoop.hive.udf.MyLength";
```

（4）在 HQL 中使用自定义的永久函数：

```
hive (default)> select ename,my_len2(ename) ename_len from emp;
```

（5）删除永久函数：

```
hive (default)> drop function my_len2;
```

7.4.2 Hive UDTF 函数

UDTF 函数用来解决输入一行、输出多行的需求。

编写 UDTF 函数需要继承 org.apache.hadoop.hive.ql.udf.generic.GenericUDTF 类，并实现 initialize()、process()、close()三个方法。UDTF 函数首先会调用 initialize()方法，此方法返回 UDTF 函数的返回行的信息（返回个数、类型）。初始化完成后，会调用 process()方法，真正的处理过程在 process()方法中。在 process()方法中，每调用一次 forward()就产生一行；如果产生多列，那么可以将多个列的值放在一个数组中，然后将该数组传入 forward()。最后调用 close()方法，对需要清理的数据进行清理。

UDTF 函数有两种使用方法，一种是直接放到 Select 后面使用，另一种是和 Lateral View 一起使用。

下面看一个 UDTF 函数的应用示例，示例使用前面的 score 表及其数据。

本示例要求自定义一个函数，能够汇总 score 表中的各科成绩。

（1）在 IDEA 中创建一个 Maven 工程 Hive：

```
groupId为com.hadoop.hiveudtf，artifactId为hiveUDTF。
```

（2）导入依赖：

```
<dependencies>
    <!-- https://mvnrepository.com/artifact/org.apache.hive/hive-exec -->
    <dependency>
        <groupId>org.apache.hive</groupId>
        <artifactId>hive-exec</artifactId>
        <version>3.1.2</version>
    </dependency>
</dependencies>
```

（3）创建一个 Java 类 Hive_UDTF：

```
package hiveUDTF;
import java.util.ArrayList;
import org.apache.hadoop.hive.ql.udf.generic.GenericUDTF;
import org.apache.hadoop.hive.ql.exec.UDFArgumentException;
import org.apache.hadoop.hive.ql.metadata.HiveException;
import org.apache.hadoop.hive.serde2.objectinspector.ObjectInspector;
import org.apache.hadoop.hive.serde2.objectinspector. ObjectInspectorFactory;
import org.apache.hadoop.hive.serde2.objectinspector.StructObjectInspector;
```

```
    import org.apache.hadoop.hive.serde2.objectinspector.primitive.
PrimitiveObjectInspectorFactory;
    public class Hive_Udtf extends GenericUDTF{
        Integer nScore = Integer.valueOf(0);
        Object forwardObj[] = new Object[1];
        String strName;
    @Override
        public StructObjectInspector initialize(ObjectInspector[] args)throws
UDFArgumentException {
            strName="";
            ArrayList<String> fieldNames = new ArrayList<String>();
            ArrayList<ObjectInspector> fieldOIs = new
ArrayList<ObjectInspector>();
            fieldNames.add("col1");

    fieldOIs.add(PrimitiveObjectInspectorFactory.javaStringObjectInspector);
            return
ObjectInspectorFactory.getStandardStructObjectInspector(fieldNames,fieldOIs);
        }
    @Override
        public void process(Object[] args) throws HiveException {
            if(!strName.isEmpty() && !strName.equals(args[0].toString())){
                //输出总分
                String[] newRes = new String[1];
                newRes[0]=(strName+"\t"+String.valueOf(nScore));
                forward(newRes);
                nScore=0;
            }
            strName=args[0].toString();
            nScore+=Integer.parseInt(args[1].toString());
        }
        @Override
        public void close() throws HiveException {
            forwardObj[0]=(strName+"\t"+String.valueOf(nScore));
            forward(forwardObj);
        }
    }
```

（4）打成 JAR 包，重命名并移动存放在/opt/datas/目录下：

```
    mv /home/hadoop/workspace/hiveUDTF/target/hiveUDTF-0.0.1-SNAPSHOT.jar
/opt/datas/hiveudtf.jar
```

（5）将 JAR 包添加到 Hive 的 classpath 目录下：

```
hive(default)>add jar /opt/datas/hiveudtf.jar;
Added [/opt/datas/hiveudtf.jar] to class path
Added resources: [/opt/datas/hiveudtf.jar]
```

（6）创建自定义函数 myUDTF：

```
hive(default)>create temporary function myUDTF  as
"hiveUDTF.Hive_Udtf";
```

（7）使用自定义函数 myUDTF 查询表 score 的数据：

```
hive(default)>select myUDTF(name,score) as sumscore from score;
```

```
OK
sum score
Steve   311
Doug     336
Linus    336
Kelly    317
```

（8）在Hive的命令行窗口中删除函数：

```
hive(default)>drop temporary function [if exists]
[dbname.]function_name;
```

第 8 章

Hive 数据压缩

Hive 查询最终是转换为 MapReduce 程序来执行的，而 MapReduce 的性能瓶颈在于网络 I/O 和磁盘 I/O，要解决性能瓶颈，最主要的方法是减少数据量，对数据进行压缩是个好的方式。虽然压缩减少了数据量，但是压缩过程需要消耗 CPU，不过 Hadoop 的性能瓶颈往往不在于 CPU，CPU 压力并不大，因此压缩可以充分利用比较空闲的 CPU。

压缩数据可以大量减少磁盘的存储空间。把存储格式和压缩数据结合使用，可以最大限度地节省存储空间。本章将主要讲解数据压缩格式、Hadoop 压缩配置、Map 输出压缩开启、Reduce 输出压缩开启、常用 Hive 表存储格式比较等内容。

8.1 数据压缩格式

在 Hive 中对中间数据或最终数据进行压缩，是提高数据性能的一种手段。压缩数据可以大量减少磁盘的占用，比如基于文本的数据文件可以压缩 40%或者更多。同时，压缩后的文件在磁盘间传输时所用的 I/O 也会大大减少。当然，压缩和解压缩会带来额外的 CPU 开销，但是可以节省更多的 I/O 操作和使用更少的内存开销。

常见的数据压缩格式有 Gzip、Bzip2、LZO、LZ4 和 Snappy 等。

可使用以下 3 种指标对压缩格式进行评价。

- 压缩比：压缩比越高，压缩后的文件越小，因此压缩比越高越好。
- 压缩速度：压缩速度越快越好。
- 可分割：已经压缩的格式文件是否可以再分割。可以分割的格式允许单一文件由多个 Map 程序处理，从而更好地并行化计算和处理。

常见的压缩格式对比如下：

- Bzip2 具有最高的压缩比，但会带来更大的 CPU 开销，Gzip 较 Bzip2 次之。如果基于磁盘利用率和 I/O 考虑，那么这两种压缩格式都比较有吸引力。

- LZO 和 Snappy 有更快的解压缩速度，如果更关注压缩、解压缩速度，那么它们都是不错的选择。LZO 和 Snappy 在压缩数据上的速度大致相当，但 Snappy 在解压缩速度上较 LZO 更快。
- Hadoop 会将大文件分割成 HDFS 块（默认 128MB）大小的分片，每个分片对应一个 Map 程序。在这几种压缩格式中，Bzip2、LZO、Snappy 压缩是可分割的，Gzip 则不支持分割。

Hive 表的存储格式与压缩格式相结合：

（1）TextFile 为默认 Hive 表存储格式，数据加载速度最快，可以采用 Gzip 进行压缩，压缩后的文件无法分割，即无法并行处理。

（2）表存储格式的压缩比最低，查询速度一般。将数据存储到 SequenceFile 格式的 Hive 表中，这时数据就会被压缩存储，并且一共有 3 种压缩格式（None、Record、Block）。SequenceFile 是一种可分割的文件格式。

（3）ORC 表存储格式的压缩比最高，查询速度最快，数据加载速度最慢。ORC 格式由于采用列式存储方式，因此数据加载时性能消耗较大，但是具有较好的压缩比和查询速度。ORC 格式支持压缩的 3 种类型为 None、Zlib、Snappy。

在 Hive 中要灵活地针对不同的应用场景使用不同的压缩方式。如果是数据源，采用 ORC+Bzip2 或 ORC+Gzip 的方式，这样可以在很大程度上节省磁盘空间；而在计算的过程中，为了不影响执行的速度，可以浪费一些磁盘空间，因此建议采用 ORC+Snappy 的方式，这样可以整体提升 Hive 的执行速度。

几种常见压缩格式比较如表 8-1 所示。

表8-1 常见压缩格式比较

压缩格式	压缩比	压缩速度	是否可分割	文件扩展名
Zlib	中	中	否	.deflate
Gzip	中	中	否	.gz
Bzip2	高	慢	是	.bz2
LZO	低	快	是	.lzo
LZ4	低	快	是	.lz4
Snappy	低	快	是	.snappy

8.2 数据压缩配置

8.2.1 Snappy 压缩方式配置

Snappy 是一个基于 C++的用来压缩和解压缩的开发包，其目标不是最大限度地压缩或者兼容其他压缩格式，而是旨在提供较高的压缩速度和合理的压缩率。Snappy 比 Zlib 更快，但文件相对要大 20%～100%。在 64 位模式的 Core i7 处理器上，可达 250～500MB/s 的压缩速度。

Snappy 的前身是 Zippy。虽然它只是一个数据压缩库，但却被 Google 用于许多内部项目，其中就包括 BigTable、MapReduce 和 RPC。Google 宣称它对这个库本身及其算法做了数据处理速度上的优化，作为代价，并没有考虑输出大小以及和其他类似工具的兼容性问题。Snappy 特地为 64 位 x86 处理器做了优化，在单个 Intel Core i7 处理器内核上能够达到至少 250MB/s 的压缩速率和 500MB/s 的解压速率。

如果允许损失一些压缩率的话，那么 Snappy 可以达到更高的压缩速度，虽然生成的压缩文件可能会比其他库的要大上 20%～100%，但是，相比其他的压缩库，Snappy 能够在特定的压缩率下拥有惊人的压缩速度，“压缩普通文本文件的速度是其他库的 1.5～1.7 倍，压缩 HTML 能达到 2～4 倍，但是对于 JPEG、PNG 以及其他的已压缩的数据，压缩速度不会有明显改善”。Snappy 从一开始就被“设计为即便遇到损坏或者恶意的输入文件都不会崩溃”，而且被 Google 在生产环境中用于压缩 PB 级的数据，其健壮性和稳定程度可见一斑。

Snappy 也可以用于和其他压缩库如 Zlib、LZO、LZF、FastLZ 和 QuickLZ 做对比测试，前提是我们在机器上安装了这些压缩库。Snappy 是一个基于 C++的库，我们可以在产品中使用，不过也有一些其他语言的版本，例如 Haskell、Java、Perl、Python 和 Ruby。

Snappy 压缩方式配置的过程如下：

（1）查看 Hadoop 命令 checknative：

```
hadoop@server201:/usr/local/hadoop$ hadoop
checknative [-a|-h]  check native hadoop and compression libraries availability
```

（2）查看 Hadoop 支持的压缩格式类型：

```
hadoop@server201:/usr/local/hadoop$ hadoop checknative
Native library checking:
hadoop:  true /usr/local/hadoop/lib/native/libhadoop.so
zlib:    true /lib64/libz.so.1
snappy:  false
lz4:     true revision:99
bzip2:   false
```

（3）导入压缩包。

将编译好的支持 Snappy 压缩的 hadoop-3.2.3.tar.gz 包导入 Linux 本地的/opt/software 目录中。

（4）解压 hadoop-3.2.3.tar.gz 到当前路径：

```
hadoop@server201:/opt/software$ tar -zxvf hadoop-3.2.3.tar.gz
```

（5）查看动态链接库。

切换到/opt/software/hadoop-3.2.3/lib/native 目录，查看支持 Snappy 压缩的动态链接库：

```
hadoop@server201 native$ pwd
/opt/software/hadoop-3.2.3/lib/native
[hadoop@server201 native]$ ll
-rw-r--r--. 1 hadoop hadoop  472950 12月1 10:18 libsnappy.a
-rwxr-xr-x. 1 hadoop hadoop   955 12月1 10:18 libsnappy.la
lrwxrwxrwx. 1 hadoop hadoop 18 12月24 20:39 libsnappy.so -> libsnappy.so.1.3.0
lrwxrwxrwx. 1 hadoop hadoop  18 12月24 20:39 libsnappy.so.1 ->
libsnappy.so.1.3.0
-rwxr-xr-x. 1 hadoop hadoop  228177 12月1 10:18 libsnappy.so.1.3.0
```

（6）动态链接库复制。

复制 /opt/software/hadoop-3.2.3/lib/native 目录中的所有动态链接库内容到 Hadoop 的 /usr/local/hadoop/lib/native 目录中：

```
hadoop@server201 native$ cp ../native/*
/usr/local/hadoop/lib/native/
```

（7）重新查看 Hadoop 支持的压缩格式类型：

```
hadoop@server201:/usr/local/hadoop$ hadoop checknative
Native library checking:
hadoop:  true /usr/local/hadoop/lib/native/libhadoop.so.1.0.0
zlib:    true /lib/x86_64-linux-gnu/libz.so.1
snappy:  true /usr/lib/x86_64-linux-gnu/libsnappy.so.1
lz4:     true revision:99
bzip2:   false
```

8.2.2 MapReduce 支持的压缩编码

Hadoop 作为一个比较通用的海量数据处理平台，在使用压缩方式方面，主要考虑压缩速度和压缩文件的可分割性。所有的压缩算法都会权衡时间和空间，更快的压缩和解压缩速度通常会耗费更多的空间（压缩比较低）。例如，通过 gzip 命令压缩数据时，用户可以设置不同的选项来选择速度优先或空间优先，选项-1 表示优先考虑速度，选项-9 表示空间最优，可以获得最高的压缩比。需要注意的是，有些压缩算法的压缩和解压缩速度会有比较大的差别，Gzip 和 Zip 是通用的压缩工具，在时间/空间处理上相对平衡，Bzip2 压缩比 Gzip 和 Zip 更有效，但速度较慢，而且 Bzip2 的解压缩速度快于它的压缩速度。

当使用 MapReduce 处理压缩文件时，需要考虑压缩文件的可分割性。考虑我们需要对保持在 HDFS 上的一个大小为 1GB 的文本文件进行处理，在当前 HDFS 的数据块大小为 64MB 的情况下，将该文件存储为 16 块，对应的 MapReduce 作业会将该文件分为 16 个输入分片，提供给 16 个独立的 Map 任务进行处理。但如果该文件是一个 Gzip 格式的压缩文件（大小不变），那么这时 MapReduce 作业不能够将该文件分为 16 个分片，因为不可能从 Gzip 数据流中的某个点开始进行数据解压。但是，如果该文件是一个 Bzip2 格式的压缩文件，那么 MapReduce 作业可以通过 Bzip2 格式压缩文件中的块，将输入划分为若干输入分片，并从块开始处解压缩数据。Bzip2 格式压缩文件中，块与块之间提供了一个 48 位的同步标记，因此，Bzip2 支持数据分割。

为了支持多种压缩和解压缩格式，Hadoop 引入了编码/解码器，如表 8-2 所示。

表8-2 压缩格式对应的编码/解码器

压缩格式	对应的编码/解码器
Zlib	org.apache.hadoop.io.compress.DefaultCodec
Gzip	org.apache.hadoop.io.compress.GzipCodec
Bzip2	org.apache.hadoop.io.compress.Bzip2Codec
LZO	com.hadoop.compression.lzo.LzopCodec
LZ4	com.hadoop.compression.lzo.Lz4Codec
Snappy	org.apache.hadoop.io.compress.SnappyCodec

MapReduce 的压缩过程主要有三个阶段：

- map 之前：要考虑数据量的大小。数据量小的就不需要考虑过多地切片，主要追求的是快速，可以选择压缩方式为 Snappy 或者 LZO。数据量大的话，考虑到切片，压缩方式可以选择 LZO 或者 Bzip2。
- map 之后：map 之后的数据要经历 reduce 聚合，主要考虑速度，因此压缩方式选择 Snappy 或者 LZO。
- reduce 之后：需要看具体的需求。类似于持久保存，那就需要看数据量的大小，数据量小的就不需要考虑过多地切片，主要追求的是快速，可以选择压缩方式为 Snappy 或者 LZO；数据量大的话，考虑到切片，压缩方式可以选择 LZO 或者 Bzip2。

8.2.3 MapReduce 压缩参数配置

若在 Hadoop 中启用压缩，则可以在 mapred-site.xml 等配置文件中配置参数，如表 8-3 所示。

表8-3 MapReduce压缩格式参数配置

参数	默认值	阶段	开启压缩建议
io.compression.codecs（在 core-site.xml 中配置）	org.apache.hadoop.io.compress.DefaultCodec, org.apache.hadoop.io.compress.GzipCodec, org.apache.hadoop.io.compress.Bzip2Codec, org.apache.hadoop.io.compress.Lz4Codec	输入压缩	Hadoop 使用文件扩展名判断是否支持某种编/解码器
mapreduce.map.output.compress	False	Map 输出	若参数设为 True，则启动压缩
mapreduce.map.output.compress.codec	org.apache.hadoop.io.compress.DefaultCodec	Map 输出	使用 LZO、LZ4 或 Snappy 编/解码器在此阶段压缩数据
mapreduce.output.fileoutputformat.compress	False	Reduce 输出	若参数设为 True，则启动压缩
mapreduce.output.fileoutputformat.compress.codec	org.apache.hadoop.io.compress.DefaultCodec	Reduce 输出	使用标准工具或者编/解码器，如 Gzip、Bzip2 和 Snappy

8.3 开启 Map 端和 Reduce 端的输出压缩

1. 开启 Map 端的输出压缩开启

开启 Map 端的输出压缩可以减少作业中 Map 任务和 Reduce 任务之间的数据传输量，具体配置参数如下：

- hive.exec.compress.intermediate：该值默认为 False，设置为 True 时激活中间数据压缩功能。
- mapreduce.map.output.compression.codec：Map 输出压缩格式的配置参数，可以使用

LZO、LZ4或Snappy编码/解码器在此阶段压缩数据。SnappyCodec压缩格式会带来很好的压缩性能和较低的CPU开销。HQL语句最终会被编译成Hadoop的MapReduce作业，开启Hive的中间数据压缩功能，就是在MapReduce的Shuffle阶段对Map产生的中间结果进行数据压缩。在这个阶段，优先选择一个低CPU开销的压缩格式。

例如：

（1）开启Hive中间传输数据压缩功能：

```
hive(hivedwh)>set hive.exec.compress.intermediate=true;
```

（2）开启MapReduce中Map输出压缩功能：

```
hive(hivedwh)>set mapreduce.map.output.compress=true;
```

（3）设置MapReduce中Map输出数据的Snappy压缩方式：

```
hive(hivedwh)>set mapreduce.map.output.compress.codec=
 org.apache.hadoop.io.compress.SnappyCodec;
```

（4）执行查询语句：

```
hive(hivedwh)>select count(ename) name from emp;
```

2. 开启Reduce端的输出压缩开启

当Hive将输出写入表中时，输出内容同样可以进行压缩。控制这个功能的属性为hive.exec.compress.output，该属性的默认值为False，这种情况下输出的就是非压缩的纯文本文件。用户可以通过在查询语句或执行脚本中设置hive.exec.compress.output的值为True，来开启输出结果压缩功能。

例如：

（1）开启Hive最终输出数据压缩功能：

```
hive(hivedwh)>set hive.exec.compress.output=true;
```

（2）开启MapReduce最终输出数据压缩功能：

```
hive(hivedwh)>set mapreduce.output.fileoutputformat.compress=true;
```

（3）设置MapReduce最终数据输出压缩方式为Snappy：

```
hive(hivedwh)>set mapreduce.output.fileoutputformat.compress.codec =
 org.apache.hadoop.io.compress.SnappyCodec;
```

（4）设置MapReduce最终数据输出压缩类型为块压缩：

```
hive(hivedwh)>set mapreduce.output.fileoutputformat.compress.type=BLOCK;
```

（5）测试输出结果是否为压缩文件：

```
hive(hivedwh)>insert overwrite local directory '/opt/datas/output'
select * from emp distribute by deptno sort by empno desc;
```

8.4 常用 Hive 表存储格式比较

Apache Hive 支持 Apache Hadoop 中常用的几种文件格式，如 TextFile、RCFile、SequenceFile、AVRO、ORC 和 Parquet 格式。Cloudera Impala 也支持这些文件格式。在建表时使用 STORED AS (TextFile|RCFile|SequenceFile|AVRO|ORC|Parquet)来指定存储格式。

- TextFile 每一行都是一条记录，每行都以换行符（\n）结尾。数据不做压缩，磁盘开销大，数据解析开销大。可结合 Gzip、Bzip2 使用（系统自动检查，执行查询时自动解压），但使用这种方式，Hive 不会对数据进行切分，从而无法对数据进行并行操作。
- SequenceFile 是 Hadoop API 提供的一种二进制文件支持，具有使用方便、可分割、可压缩的特点。支持三种压缩选择：None、Record、Block。Record 压缩率低，一般建议使用 Block 压缩。
- RCFile 是一种行列存储相结合的存储方式。首先，它将数据按行分块，保证同一个 record（记录）在同一个块上，避免读一个记录需要读取多个块。其次，块数据列式存储，有利于数据压缩和快速列存取。
- AVRO 是开源项目，为 Hadoop 提供数据序列化和数据交换服务。我们可以在 Hadoop 生态系统和以任何编程语言编写的程序之间交换数据。Avro 是基于大数据 Hadoop 的应用程序中流行的文件格式之一。
- ORC 文件代表了优化排柱状的文件格式，提供了一种将数据存储在 Hive 表中的高效方法。这个文件系统实际上是为了克服其他 Hive 文件格式的限制而设计的。在 Hive 从大型表读取、写入和处理数据时，使用 ORC 文件可以提高性能。
- Parquet 是一个面向列的二进制文件格式，对于大型查询的类型是高效的。对于扫描特定表格中的特定列的查询，Parquet 特别有用。Parquet 一般使用 Snappy、Gzip 压缩，目前默认为 Snappy。

各种存储格式的对比如表 8-4 所示。

表8-4 存储格式的对比

存储格式	存储方式	特点
TextFile	行存储	存储空间消耗比较大，并且压缩的 text 无法分割和合并，查询的效率最低，可以直接存储，加载数据的速度最快
SequenceFile	行存储	存储空间消耗最大，压缩的文件可以分割和合并，查询效率高，需要通过 text 文件转换来加载
RCFile	数据按行分块每块按照列存储	存储空间最小，查询的效率最高，需要通过 text 文件转换来加载，加载的速度最低。 压缩快，快速列存取。 读记录涉及的 block 最少。 读取需要的列时只需要读取每个 row group 的头部定义。 读取全量数据的操作性能相比 SequenceFile 可能没有明显的优势

（续表）

存储格式	存储方式	特　点
ORCFile	数据按行分块每块按照列存储	压缩快，快速列存取，效率比 RCFile 高，是 RCFile 的改良版本
Parquet	列存储	相对于 PRC，Parquet 压缩比较低，查询效率较低，不支持 update、insert 和 ACID，但是 Parquet 支持 Impala 查询引擎

Parquet 和 ORC 的对比如表 8-5 所示。

表8-5　Parquet与ORC的对比

对比项	Parquet	ORC
网站	http://parquet.apache.org	http://orc.apache.org
发展状态	目前是 Apache 开源的顶级项目，列式存储引擎	目前是 Apache 开源的顶级项目，列式存储引擎
开发语言	Java	Java
主导公司	Twitter/Cloudera	Hortonworks
ACID	不支持	支持 ACID 事务
修改操作（update、delete）	不支持	支持
支持索引（统计信息）	粗粒度索引， block/group/chunk 级别统计信息	粗粒度索引， file/stripe/row 级别统计信息， 不能精确到列建立索引
支持的查询引擎	Apache Drill、Impala	Apache Hive
查询性能	ORC 性能更高一点	ORC 性能更高一点
压缩比	ORC 压缩比更高	ORC 压缩比更高
列编码	支持多种编码，如字典、RLE、Delta 等	支持主流编码，与 Parquet 类似

下面从压缩比和查询速度两个方面来测试对比几种常见的 Hive 表存储格式。

用于测试的数据为一个 41.3MB 大小的用户文件，其中共有 2139109 条记录数据，每条记录有 3 个字段。

1. TextFile 格式

（1）创建表，存储数据格式为 TextFile：

```
hive(hivedwh)>create table text_user(
userid string,
view int,
click int)
row format delimited fields terminated by '\t'
stored as textfile;
```

（2）向表中加载数据：

```
hive(hivedwh)>load data local inpath '/opt/datas/user.txt' into table text_ user;
```

（3）查看表中数据大小：

```
hive(hivedwh)>dfs -du -h /user/hive/warehouse/hivedwh.db/text_user;
34.8 M  /user/hive/warehouse/hivedwh.db/text_user/user.txt
```

2. SequenceFile 格式

（1）创建表，存储数据格式为 SequenceFile：

```
hive(hivedwh)>create table seq_user(
userid string,
view int,
click int)
row format delimited fields terminated by '\t'
stored as SequenceFile;
```

（2）向表中加载数据：

```
hive(hivedwh)>insert into table seq_user select * from text_user;
```

（3）查看表中数据大小：

```
hive(hivedwh)>dfs -du -h /user/hive/warehouse/hivedwh.db/seq_user;
57.8 M  /user/hive/warehouse/hivedwh.db/seq_user/000000_0
```

3. ORC 格式

（1）创建表，存储数据格式为 ORC：

```
hive(hivedwh)>create table orc_user(
userid string,
view int,
click int)
row format delimited fields terminated by '\t'
stored as orc;
```

（2）向表中加载数据：

```
hive(hivedwh)>insert into table orc_user select * from text_user;
```

（3）查看表中数据大小：

```
hive(hivedwh)>dfs -du -h /user/hive/warehouse/hivedwh.db/orc_user;
17.3 M  /user/hive/warehouse/hivedwh.db/orc_user/000000_0
```

4. Parquet 格式

（1）创建表，存储数据格式为 Parquet：

```
hive(hivedwh)>create table par_user (
userid string,
view int,
click int)
row format delimited fields terminated by '\t'
stored as parquet;
```

（2）向表中加载数据：

```
hive(hivedwh)>insert into table par_user select * from text_user;
```

（3）查看表中数据大小：

```
hive(hivedwh)>dfs -du -h /user/hive/warehouse/hivedwh.db/par_user;
35.1 M  /user/hive/warehouse/hivedwh.db/par_user/000000_0
```

5. 常见的 Hive 表存储格式的压缩比测试总结

经过测试，上面所讲解的 4 种 Hive 表存储格式的压缩比数据如表 8-6 所示。注意：表中压缩比计算公式为：

压缩比=（原始数据－压缩后数据）/原始数据

表8-6 几种常见的Hive表存储格式的压缩比测试结果

压缩格式	TextFile	SequenceFile	ORC	Parquet
数据量（MB）	34.8	57.8	17.3	35.1
压缩比（%）	0	-66	50	-1

从表 8-4 可以看出，几种常见的 Hive 表存储格式的压缩比从大到小的顺序为 ORC>TextFile>Parquet>SequenceFile。这个结果适用于现有的数据量，在数据量较小的情况下，可能会出现与之相悖的结论。

6. 存储文件的查询速度测试

（1）TextFile 格式：

```
hive(hivedwh)>select count(*) count_text from text_user;
OK
count_text
2139109
Time taken: 22.089 seconds, Fetched: 1 row(s)
```

（2）SequenceFile 格式：

```
hive(hivedwh)>select count(*) count_seq from seq_user;
OK
count_seq
2139109
Time taken: 24.153 seconds, Fetched: 1 row(s)
```

（3）ORC 格式：

```
hive(hivedwh)>select count(*) count_orc from orc_user;
OK
count_orc
2139109
Time taken: 20.698 seconds, Fetched: 1 row(s)
```

（4）Parquet 格式：

```
hive(hivedwh)>select count(*) count_par from par_user;
OK
count_par
2139109
Time taken: 21.457 seconds, Fetched: 1 row(s)
```

（5）几种常见的 Hive 表存储格式的查询速度测试总结，如表 8-7 所示。

表8-7 常见的Hive表存储格式的查询时间测试总结

压缩格式	TextFile	SequenceFile	ORC	Parquet
查询时间（s）	22.089	24.153	20.689	21.457

从表 8-5 可以看出，SequenceFile 格式查询时间略长，其他格式无明显差异。也就是说，这 4 种存储格式的查询速度比较接近。这个结果适用于当前现有的数据量，在数据量较小的情况下，可能会出现与之相悖的结论。

8.5 存储与压缩相结合

Hive 表 ORC 存储格式的压缩支持 3 种类型，分别为 None、Zlib、Snappy，其中 Zlib 为默认类型。ORC 存储格式的压缩如表 8-8 所示。

表8-8 ORC存储格式的压缩

key	默认值	描述
orc.compress	Zlib	支持压缩的 3 种类型：None、Zlib、Snappy
orc.compress.size	262 144	每个压缩 Chunk 的字节大小
orc.stripe.size	67 108 864	每个 Chunk 的字节大小
orc.row.index.stride	10 000	stride 中的行数，即分组大小不低于 1000
orc.create.index	True	是否需要创建行索引
orc.bloom.filter.columns	""	列分隔符
orc.bloom.filter.fpp	0.05	bloom filter 的概率为 0.0～1.0

下面分别创建 ORC 存储格式的表，并修改其数据压缩类型。导入数据后，比较其压缩比数据。

1. 创建 Zlib 压缩类型的 ORC 存储格式表

（1）创建表：

```
hive(hivedwh)>create table orc_zlib(
userid string,
view int,
click int)
row format delimited fields terminated by '\t'
stored as orc;
```

（2）设置数据压缩格式：

```
hive(hivedwh)>alter table orc_zlib set
tblproperties("orc.compress"="ZLIB");
```

（3）查看数据压缩格式：

```
hive(hivedwh)>desc formatted orc_zlib;
orc.compress        ZLIB
```

（4）向 Hive 表导入数据：

```
hive(hivedwh)>insert into table orc_zlib select * from text_user;
```

2. 创建 None 压缩类型的 ORC 存储格式表

（1）创建表：

```
hive(hivedwh)>create table orc_none(
userid string,
view int,
click int)
row format delimited fields terminated by '\t'
stored as orc;
```

（2）设置数据压缩格式：

```
hive(hivedwh)>alter table orc_none set
tblproperties("orc.compress"="NONE");
```

（3）查看数据压缩格式：

```
hive(hivedwh)>desc formatted orc_none;
orc.compress          NONE
```

（4）向 Hive 表导入数据：

```
hive(hivedwh)>insert into table orc_none select * from text_user;
```

3. 创建 Snappy 压缩类型的 ORC 存储格式表

（1）创建表：

```
hive(hivedwh)>create table orc_snappy(
userid string,
view int,
click int)
row format delimited fields terminated by '\t'
stored as orc;
```

（2）设置数据压缩格式：

```
hive(hivedwh)>alter table orc_snappy set
tblproperties("orc.compress"="SNAPPY");
```

（3）查看数据压缩格式：

```
hive(hivedwh)>desc formatted orc_snappy;
orc.compress          SNAPPY
```

（4）向 Hive 表导入数据：

```
hive(hivedwh)>insert into table orc_snappy select * from text_user;
```

4. 查看加载后数据

```
hive(hivedwh)>dfs -du -h /user/hive/warehouse/hivedwh.db/orc*;
27.1 M  /user/hive/warehouse/hivedwh.db/orc_none/000000_0
25.2 M  /user/hive/warehouse/hivedwh.db/orc_snappy/000000_0
17.3 M  /user/hive/warehouse/hivedwh.db/orc_user/000000_0
17.3 M  /user/hive/warehouse/hivedwh.db/orc_zlib/000000_0
```

这种 ORC 存储格式文件比 Snappy 压缩的文件还要小。原因是 ORC 存储格式文件默认采用 Zlib 压缩，比 Snappy 压缩的小。在 ORC 存储格式中，None 类型实际上对原始数据也做了压缩。从现有数据看，数据压缩比从大到小的顺序依次为 Zlib、Snappy 和 None 类型。

5. 存储格式和压缩类型相结合的总结

存储格式和压缩类型相结合，能够产生更好的效果。在实际的项目开发中，Hive 表的文件存储格式一般选择 ORC 或 Parquet 格式。压缩类型一般选择 Snappy、LZO。这样一方面减少磁盘的使用量，另一方面可以实现数据的 Split（分布式计算），使得查询速度更快。

第 9 章

Hive 调优

Hive 优化需要结合业务需求、数据的类型、分布、质量等实际状况，综合考虑如何进行系统性的优化。Hive 底层是 MapReduce，因此 Hadoop 优化是 Hive 优化的基础。Hive 优化包括 Hive 参数优化、数据倾斜的解决、HQL 优化等方面。

9.1 Hadoop 计算框架特性

Hadoop 计算框架具有如下特性：

- 数据量大不是问题，数据倾斜是个问题。
- job 数比较多的作业运行效率相对比较低，比如即使只有几百行的表，单如果多次关联多次汇总，产生十几个 job，那么耗时会很长，原因是 MapReduce 作业初始化的时间是比较长的。
- sum、count、max、min 等 UDAF 不怕数据倾斜问题，因为 Hadoop 在 Map 端的汇总合并优化使数据倾斜不成问题。
- count(distinct)在数据量大的情况下效率较低，如果是多 count(distinct)，效率更低。因为 count(distinct)是按 group by 字段分组，按 distinct 字段排序，一般这种分布方式是很倾斜的。比如淘宝一天 30 亿的 PV(Page View)，如果按性别分组，分配两个 Reduce，那么每个 Reduce 要处理 15 亿数据。

9.2 Hive 优化的常用手段

Hive 优化的常用手段如下：

- 好的模型设计事半功倍。
- 解决数据倾斜问题。
- 减少 job 数。
- 设置合理的 MapReduce 的 task 数，能有效提升性能。比如，10w+级别的计算，用 160 个 Reduce，那是相当的浪费，设置 1 个足够了。
- 了解数据分布，自己动手解决数据倾斜问题是个不错的选择。set hive.groupby.skewindata =true 是通用的算法优化，但算法优化有时不能适应特定业务背景，因此开发人员需要了解业务、了解数据，通过业务逻辑精确有效地解决数据倾斜问题。
- 数据量较大的情况下，慎用 count(distinct)，count(distinct)容易产生数据倾斜问题。
- 对小文件进行合并，这是行之有效的提高调度效率的方法。假如所有的作业都设置合理的文件数，那它对任务的整体调度效率也会产生积极的正向影响。
- 优化时把握整体，单个作业最优不如整体最优。

9.3 Hive 优化要点

9.3.1 全排序

Hive 的排序关键字是 SORT BY，它有意区别于传统数据库的 ORDER BY 也是为了强调两者的区别——SORT BY 只能在单机范围内排序。下面看两个示例。

示例 1：以下为网站访问日志记录数据表 c02_clickstat_fatdt1，指定了两个 Reduce 进行数据分发（建表语句、日志文件参见配套源码）。

```
set mapred.reduce.tasks=2;
```

原值：

```
    hive> select cookie_id,page_id,id from c02_clickstat_fatdt1
    where cookie_id
IN('1.193.131.218.1288611279693.0','1.193.148.164.1288609861509.2')
    1.193.148.164.1288609861509.2   113181412886099008861288609901078194082403
684000005
    1.193.148.164.1288609861509.2   127001128860563972141288609859828580660473
684000015
    1.193.148.164.1288609861509.2   113181412886099165721288609915890452725326
684000018
    1.193.131.218.1288611279693.0   01c183da6e4bc50712881288611540109914561053
684000114
    1.193.131.218.1288611279693.0   01c183da6e4bc22412881288611414343558274174
684000118
    1.193.131.218.1288611279693.0   01c183da6e4bc50712881288611511781996667988
684000121
    1.193.131.218.1288611279693.0   01c183da6e4bc22412881288611523640691739999
684000126
    1.193.131.218.1288611279693.0   01c183da6e4bc50712881288611540109914561053
684000128
```

SORT 排序后的值：

```
    hive> select cookie_id,page_id,id from c02_clickstat_fatdt1 where
    cookie_id IN('1.193.131.218.1288611279693.0','1.193.148.164.1288609861509.2')
    SORT BY COOKIE_ID,PAGE_ID;
    1.193.131.218.1288611279693.0           684000118
01c183da6e4bc22412881288611414343558274174      684000118
    1.193.131.218.1288611279693.0           684000114
01c183da6e4bc50712881288611540109914561053      684000114
    1.193.131.218.1288611279693.0           684000128
01c183da6e4bc50712881288611540109914561053      684000128
    1.193.148.164.1288609861509.2           684000005
113181412886099008861288609901078194082403      684000005
    1.193.148.164.1288609861509.2           684000018
113181412886099165721288609915890452725326      684000018
    1.193.131.218.1288611279693.0           684000126
01c183da6e4bc22412881288611523640691739999      684000126
    1.193.131.218.1288611279693.0           684000121
01c183da6e4bc50712881288611511781996667988      684000121
    1.193.148.164.1288609861509.2           684000015
127001128860563972141288609859828580660473      684000015
```

ORDER 排序后的值：

```
    hive> select cookie_id,page_id,id from c02_clickstat_fatdt1
    where cookie_id
IN('1.193.131.218.1288611279693.0','1.193.148.164.1288609861509.2')
    ORDER BY PAGE_ID,COOKIE_ID;
    1.193.131.218.1288611279693.0           684000118
01c183da6e4bc22412881288611414343558274174      684000118
    1.193.131.218.1288611279693.0           684000126
01c183da6e4bc22412881288611523640691739999      684000126
    1.193.131.218.1288611279693.0           684000121
01c183da6e4bc50712881288611511781996667988      684000121
    1.193.131.218.1288611279693.0           684000114
01c183da6e4bc50712881288611540109914561053      684000114
    1.193.131.218.1288611279693.0           684000128
01c183da6e4bc50712881288611540109914561053      684000128
    1.193.148.164.1288609861509.2           684000005
113181412886099008861288609901078194082403      684000005
    1.193.148.164.1288609861509.2           684000018
113181412886099165721288609915890452725326      684000018
    1.193.148.164.1288609861509.2           684000015
127001128860563972141288609859828580660473      684000015
```

可以看到 SORT 和 ORDER 排序出来的值不一样。结果不一样的主要原因是上述查询没有Reduce key，Hive会生成随机数作为Reduce key。这样的话输入记录也随机地被分发到不同的Reducer机器上去了。为了保证Reducer之间没有重复的cookie_id记录，可以使用DISTRIBUTE BY关键字指定分发key为cookie_id。

```
    hive> select cookie_id,country,id,page_id,id from c02_clickstat_fatdt1 where
cookie_id IN('1.193.131.218.1288611279693.0','1.193.148.164.1288609861509.2')
distribute by cookie_id SORT BY COOKIE_ID,page_id;
    1.193.131.218.1288611279693.0           684000118
```

```
01c183da6e4bc22412881288611414343558274174      684000118
   1.193.131.218.1288611279693.0         684000126
01c183da6e4bc22412881288611523640691739999      684000126
   1.193.131.218.1288611279693.0         684000121
01c183da6e4bc50712881288611511781996667988      684000121
   1.193.131.218.1288611279693.0         684000114
01c183da6e4bc50712881288611540109914561053      684000114
   1.193.131.218.1288611279693.0         684000128
01c183da6e4bc50712881288611540109914561053      684000128
   1.193.148.164.1288609861509.2         684000005
11318141288609900886128860990107819408240 3      684000005
   1.193.148.164.1288609861509.2         684000018
113181412886099165721288609915890452725326      684000018
   1.193.148.164.1288609861509.2         684000015
127001128860563972141288609859828580660473      684000015
```

示例 2：使用 DISTRIBUTE BY 关键字指定分发 key。

```
CREATE TABLE if not exists t_order(
id int, -- 订单编号
sale_id int, -- 销售 ID
customer_id int, -- 客户 ID
product _id int, -- 产品 ID
amount int -- 数量
) PARTITIONED BY (ds STRING);
```

在表中查询所有销售记录，并按照销售 ID 和销售数量排序：

```
set mapred.reduce.tasks=2;
Select sale_id, amount from t_order
Sort by sale_id, amount;
```

这一查询可能得到非期望的排序。指定的两个 Reducer 分发到的数据可能是（各自排序）：

```
Reducer1:
Sale_id | amount
0 | 100
1 | 30
1 | 50
2 | 20
Reducer2:
Sale_id | amount
0 | 110
0 | 120
3 | 50
4 | 20
```

使用 DISTRIBUTE BY 关键字指定分发 key 为 sale_id，改造后的 HQL 如下：

```
set mapred.reduce.tasks=2;
Select sale_id, amount from t_order
Distribute by sale_id
Sort by sale_id, amount;
```

这样能够保证查询的销售记录集合中，销售 ID 对应的数量是正确排序的，但是销售 ID 不能正确排序，原因是Hive使用Hadoop默认的HashPartitioner分发数据。

这就涉及一个全排序的问题。解决的办法无外乎两种：

（1）不分发数据，使用单个Reducer：

```
set mapred.reduce.tasks=1;
```

这一方法的缺陷在于Reduce端成为了性能瓶颈，而且在数据量大的情况下一般都无法得到结果。但是实践中这仍然是最常用的方法，原因是通常排序的查询是为了得到排名靠前的若干结果，因此可以用limit子句大大减少数据量。使用limit n后，传输到Reduce端（单机）的数据记录数就减少到n*(map个数)。

（2）修改Partitioner，这种方法可以做到全排序。这里可以使用Hadoop自带的TotalOrderPartitioner（来自Yahoo!的TeraSort项目），这是一个为了支持跨Reducer分发有序数据而开发的Partitioner，它需要一个SequenceFile格式的文件指定分发的数据区间。如果我们已经生成了这一文件（存储在/tmp/range_key_list，分成100个Reducer），那么可以将上述查询改写为：

```
    set mapred.reduce.tasks=100;
    set hive.mapred.partitioner=org.apache.hadoop.mapred.lib.
TotalOrderPartitioner;
    set total.order.partitioner.path=/tmp/ range_key_list;
    Select sale_id, amount from t_order
    Cluster by sale_id
    Sort by amount;
```

有很多种方法生成这一区间文件（例如Hadoop自带的o.a.h.mapreduce.lib.partition.InputSampler工具），这里介绍用Hive生成的方法，例如一个id有序的t_sale表：

```
CREATE TABLE if not exists t_sale (
id int,
name string,
loc string
);
```

生成按sale_id分发的区间文件的方法是：

```
create external table range_keys(sale_id int)
row format serde
'org.apache.hadoop.hive.serde2.binarysortable.BinarySortableSerDe'
stored as
inputformat
'org.apache.hadoop.mapred.TextInputFormat'
outputformat
'org.apache.hadoop.hive.ql.io.HiveNullValueSequenceFileOutputFormat'
location '/tmp/range_key_list';
insert overwrite table range_keys
select distinct sale_id
from source t_sale sampletable(BUCKET 100 OUT OF 100 ON rand()) s
sort by sale_id;
```

生成的文件(/tmp/range_key_list目录下)可以让TotalOrderPartitioner按sale_id有序地分发Reduce处理的数据。区间文件需要考虑的主要问题是数据分发的均衡性，这有赖于对数据的深入理解。

9.3.2 怎样做笛卡儿积

当 Hive 设定为严格模式（hive.mapred.mode=strict）时，不允许在 HQL 语句中出现笛卡儿积，这实际说明了 Hive 对笛卡儿积的支持较弱。因为找不到 Join key，所以 Hive 只能使用 1 个 Reducer 来完成笛卡儿积。

当然也可以用 limit 的办法来减少某个表参与 join 的数据量，但对于需要笛卡儿积语义的需求来说，通常是一张大表和一张小表的 Join 操作，结果仍然很大（以至于无法用单机处理），这时 MapJoin 才是最好的解决办法。

MapJoin，顾名思义，会在 Map 端完成 Join 操作。这需要将 Join 操作的一张或多张表完全读入内存。

MapJoin 的用法是在查询/子查询的 SELECT 关键字后面添加“/*+ MAPJOIN(tablelist) */”提示优化器转化为 MapJoin（目前 Hive 的优化器不能自动优化 MapJoin）。其中 tablelist 可以是一张表，也可以是以逗号连接的表的列表。tablelist 中的表将会读入内存，应该将小表写在这里。

另外，有用户说 MapJoin 在子查询中可能出现未知 BUG。在大表和小表做笛卡儿积时，规避笛卡儿积的方法是，给 Join 添加一个 Join key，原理很简单：将小表扩充一列 Join key，并将小表的条目复制数倍，Join key 各不相同；将大表扩充一列 Join key，且为随机数。

9.3.3 怎样写 exist/in 子句

Hive 不支持 where 子句中的子查询，因此 SQL 常用的 exist/in 子句需要改写。这一改写相对简单。考虑以下 SQL 查询语句：

```
SELECT a.key, a.value
FROM a
WHERE a.key in
(SELECT b.key
FROM B);
```

可以改写为：

```
SELECT a.key, a.value
FROM a LEFT OUTER JOIN b ON (a.key = b.key)
WHERE b.key <> NULL;
```

一个更高效的实现是利用 LEFT SEMI JOIN 改写为：

```
SELECT a.key, a.val
FROM a LEFT SEMI JOIN b on (a.key = b.key);
```

LEFT SEMI JOIN 是 Hive 0.5.0 以上版本的特性。

9.3.4 怎样决定 Reducer 个数

Hadoop MapReduce 程序中，Reducer 个数的设置极大地影响了执行效率，这使得 Hive 怎样决定 Reducer 个数成为一个关键问题。遗憾的是 Hive 的估计机制很弱，不指定 Reducer 个数的情况下，Hive 会基于以下两个设定通过猜测来确定一个 Reducer 个数：

- hive.exec.reducers.bytes.per.reducer：默认为 1000^3。
- hive.exec.reducers.max：默认为 999。

计算 Reducer 数的公式为：

N=min(参数 2，总输入数据量/参数 1)

通常情况下，有必要手动指定 Reducer 个数。考虑到 Map 阶段的输出数据量通常会比输入有大幅减少，因此即使不设定 Reducer 个数，重设参数 2 还是必要的。依据 Hadoop 的经验，可以将参数 2 设定为 0.95*（集群中 TaskTracker 个数）。

9.3.5 合并 MapReduce 操作

1. Multi-group by

Multi-group by 是 Hive 的一个非常好的特性，它使得在 Hive 中利用中间结果变得非常方便。例如：

```
FROM (SELECT a.status, b.school, b.gender
FROM status_updates a JOIN profiles b
ON (a.userid = b.userid and
a.ds='2009-03-20' )
) subq1
INSERT OVERWRITE TABLE gender_summary
PARTITION(ds='2009-03-20')
SELECT subq1.gender, COUNT(1) GROUP BY subq1.gender
INSERT OVERWRITE TABLE school_summary
PARTITION(ds='2009-03-20')
SELECT subq1.school, COUNT(1) GROUP BY subq1.school
```

上述查询语句使用了 Multi-group by 特性连续 group by 了两次数据，使用了不同的 group by key。这一特性可以减少一次 MapReduce 操作。

2. Multi-distinct

Multi-distinct 是淘宝开发的另一个 multi-xxx 特性，使用 Multi-distinct 可以在同一查询/子查询中使用多个 distinct，这就减少了多次 MapReduce 操作

9.3.6 Bucket 与 Sampling

Bucket 是指将数据以指定列的值为 key 进行 hash，并 hash 到指定数目的桶中。这样就可以支持高效采样了。

下例就是将 userid 这一列作为 Bucket 的依据，共设置 32 个 Bucket：

```
CREATE TABLE page_view(viewTime INT, userid BIGINT,
                page_url STRING, referrer_url STRING,
                ip STRING COMMENT 'IP Address of the User')
    COMMENT 'This is the page view table'
    PARTITIONED BY(dt STRING, country STRING)
    CLUSTERED BY(userid) SORTED BY(viewTime) INTO 32 BUCKETS
    ROW FORMAT DELIMITED
```

```
            FIELDS TERMINATED BY '1'
            COLLECTION ITEMS TERMINATED BY '2'
            MAP KEYS TERMINATED BY '3'
     STORED AS SEQUENCEFILE;
```

Sampling 可以在全体数据上进行采样，这样效率自然就低，它还是要去访问所有数据。而如果一张表已经对某一列制作了 Bucket，就可以采样所有桶中指定序号的某个桶，这就减少了访问量。

下例就是采样 page_view 的 32 个桶中的第 3 个桶：

```
SELECT * FROM page_view TABLESAMPLE(BUCKET 3 OUT OF 32);
```

9.3.7 Partition

Partition 就是分区。分区通过在创建表时启用 partition by 实现，用来 partition 的维度并不是实际数据的某一列，具体分区的标志是在插入内容时给定的。当要查询某一分区的内容时可以采用 where 语句，形似 where tablename.partition_key > a 来实现。

（1）创建含分区的表：

```
CREATE TABLE page_view(viewTime INT, userid BIGINT,
             page_url STRING, referrer_url STRING,
             ip STRING COMMENT 'IP Address of the User')
PARTITIONED BY(date STRING, country STRING)
ROW FORMAT DELIMITED
      FIELDS TERMINATED BY '1'
STORED AS TEXTFILE;
```

（2）载入内容，并指定分区标志：

```
LOAD DATA LOCAL INPATH '/tmp/pv_2008-06-08_us.txt' INTO TABLE page_view
PARTITION(date='2008-06-08', country='US');
```

（3）查询指定标志的分区内容：

```
SELECT page_views.*
   FROM page_views
   WHERE page_views.date >= '2008-03-01' AND page_views.date <= '2008-03-31'
AND page_views.referrer_url like '%xyz.com';
```

9.3.8 Join

1. Join 原则

在使用写有 Join 操作的查询语句时有一条原则：应该将条目少的表/子查询放在 Join 操作符的左边。原因是在 Join 操作的 Reduce 阶段，位于 Join 操作符左边的表的内容会被加载进内存。将条目少的表放在左边，可以有效减少发生 OOM（Out of Memory，内存溢出）错误的概率。对于一条语句中有多个 Join 的情况：

（1）如果 Join 的条件相同，比如下面的查询：

```
INSERT OVERWRITE TABLE pv_users
  SELECT pv.pageid, u.age FROM page_view p
  JOIN user u ON (pv.userid = u.userid)
```

```
  JOIN newuser x ON (u.userid = x.userid);
```

- 如果 Join 的 key 相同，那么不管有多少张表，最后都会合并为一个 MapReduce。
- 一个 MapReduce 任务，而不是 n 个。
- 在做 OUTER JOIN 的时候也是一样 。

（2）如果 Join 的条件不相同，比如：

```
INSERT OVERWRITE TABLE pv_users
  SELECT pv.pageid, u.age FROM page_view p
  JOIN user u ON (pv.userid = u.userid)
  JOIN newuser x on (u.age = x.age);
```

MapReduce 的任务数目和 Join 操作的数目是对应的，上述查询和以下查询是等价的：

```
INSERT OVERWRITE TABLE tmptable
  SELECT * FROM page_view p JOIN user u
  ON (pv.userid = u.userid);
INSERT OVERWRITE TABLE pv_users
  SELECT x.pageid, x.age FROM tmptable x
  JOIN newuser y ON (x.age = y.age);
```

2. MapJoin

Join 操作在 Map 阶段完成，不再需要 Reduce，前提条件是需要的数据在 Map 过程中可以访问到。比如下面的查询：

```
INSERT OVERWRITE TABLE pv_users
  SELECT /*+ MAPJOIN(pv) */ pv.pageid, u.age
  FROM page_view pv
    JOIN user u ON (pv.userid = u.userid);
```

可以在 Map 阶段完成 Join，如图 9-1 所示。

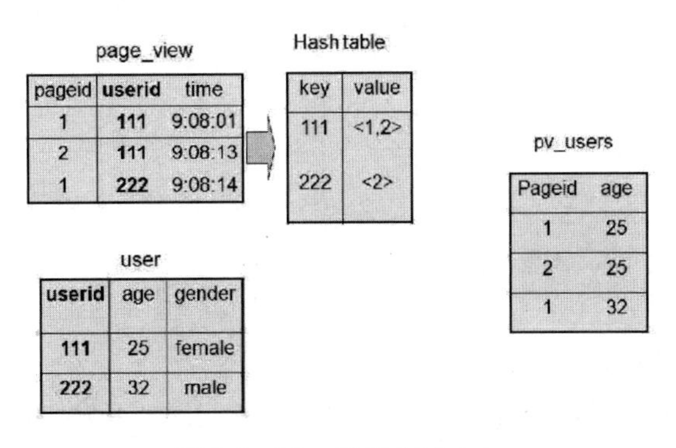

图 9-1 Map 阶段完成 Join

相关的参数为：

- hive.join.emit.interval=1000，是在发出 Join 结果之前对 Join 最右操作缓存多少行的设定。
- hive.mapjoin.size.key=10000。
- hive.mapjoin.cache.numrows=10000。

9.3.9 数据倾斜

1. 空值数据倾斜

场景：日志中常有信息丢失的问题，比如全网日志中的 user_id，如果取其中的 user_id 和 bmw_users 关联，就会碰到数据倾斜的问题。

解决方法 1：user_id 为空的不参与关联。

```
Select * From log a
Join bmw_users b
On a.user_id is not null
And a.user_id = b.user_id
Union all
Select * from log a
where a.user_id is null;
```

解决方法 2：把空值的 key 变成一个字符串加上随机数。

```
Select *
from log a
left outer join bmw_users b
on case when a.user_id is null then concat('dp_hive',rand() ) else a.user_id
end = b.user_id;
```

结论：解决方法 2 比解决方法 1 效率更好，不但 I/O 少了，而且作业数也少了。方法 1 中 log 读取两次，job 数是 2；方法 2 的 job 数是 1。这个优化适合无效 id（比如-99、''、null 等）产生的倾斜问题。把空值的 key 变成一个字符串加上随机数，就能把倾斜的数据分到不同的 Reduce 上，以此解决数据倾斜问题。

2. 不同数据类型关联产生数据倾斜

场景：一张表 s8 的日志，每个商品一条记录，要和商品表关联，但关联却碰到倾斜的问题。s8 的日志中有字符串的商品 id（类型是 string 的），也有数字的商品 id（数字 id 是 bigint 的）。我们猜测倾斜的原因是把 s8 的字符串商品 id 转成数字商品 id 做 hash 来分配 Reduce，因此字符串 id 的 s8 日志都到一个 Reduce 上了，解决方法验证了这个猜测。

解决方法：把数字类型转换成字符串类型。

```
Select * from s8_log a
Left outer join r_auction_auctions b
On a.auction_id = cast(b.auction_id as string);
```

3. 大表 Join 的数据偏斜

MapReduce 编程模型下开发代码需要考虑数据偏斜的问题，开发 Hive 代码也一样要考虑。数据偏斜的情况有以下两种：

- 情况 1：Map 输出的 key 数量极少，导致 Reduce 端（Reducer）退化为单机作业。
- 情况 2：Map 输出的 key 分布不均，少量 key 对应大量 value，导致 Reduce 端单机瓶颈。

Hive 中使用 MapJoin 解决数据偏斜的问题，即将其中的某张表（全量）分发到所有 Map 端进行 Join，从而避免了 Reduce。这要求分发的表可以被全量载入内存。

极限情况下，Join 两边的表都是大表，就无法使用 MapJoin。这种问题最为棘手，目前已知的解决思路有两种：

（1）如果是上述情况 1，那么考虑先对 Join 中的一张表去重，以此结果过滤无用信息。这样一般会将其中一张大表转化为小表，再使用 MapJoin。

一个实例是广告投放效果分析，将广告投放者信息表 i 中的信息填充到广告曝光日志表 w 中，使用投放者 id 关联。因为实际广告投放者数量很少（但是投放者信息表 i 很大），所以可以考虑先在 w 表中去重查询所有实际广告投放者 id 列表，以此 Join 过滤表 i，这一结果必然是一个小表，就可以使用 MapJoin 了。

（2）如果是上述情况 2，考虑切分 Join 中的一张表为多片，以便将切片全部载入内存，然后采用多次 MapJoin 得到结果。

一个实例是商品浏览日志分析，将商品信息表 i 中的信息填充到商品浏览日志表 w 中，使用商品 id 关联，但是某些热卖商品浏览量很大，造成数据偏斜。例如，以下语句实现了一个 inner join 逻辑，将商品信息表拆分成两张表：

```
select * from
(
select w.id, w.time, w.amount, i1.name, i1.loc, i1.cat
from w left outer join i sampletable(1 out of 2 on id) i1
)
union all
(
select w.id, w.time, w.amount, i2.name, i2.loc, i2.cat
from w left outer join i sampletable(1 out of 2 on id) i2
)
);
```

以下语句实现了 left outer join 逻辑：

```
select t1.id, t1.time, t1.amount,
   coalease(t1.name, t2.name),
   coalease(t1.loc, t2.loc),
   coalease(t1.cat, t2.cat)
from (
   select w.id, w.time, w.amount, i1.name, i1.loc, i1.cat
   from w left outer join i sampletable(1 out of 2 on id) i1
) t1 left outer join i sampletable(2 out of 2 on id) t2;
```

上述语句使用 Hive 的 sample table 特性对表进行切分。

9.3.10 合并小文件

1. 小文件产生的原因及其影响

小文件产生的原因如下：

- 动态分区使用不合理，动态分区插入的时候，误判导致产生大量分区，大量分区自然就产生大量小文件。
- Reduce 的数量设置较多，到 Reduce 处理时，会分配到不同的 Reduce 中，导致产生大量的小文件。

- 源数据文件本身就存在大量的小文件。

小文件过多产生的影响：

- 对底层存储HDFS文件系统来讲，HDFS本身就不适合存储大量小文件，小文件过多会导致NameNode元数据特别大，占用太多内存，严重影响HDFS的性能。
- 对Hive来讲，在进行查询时，每一个小文件都会被当作一个块，启动一个Map Task来完成，而一个Map Task的启动和初始化的时间远远大于逻辑处理的时间，就会形成很大的资源浪费，并且同时可执行的Map Task数量也是受限的。

文件数目过多，会给HDFS带来压力，影响处理效率，可以通过合并Map和Reduce的结果文件来消除这样的影响：

- hive.merge.mapfiles = true：是否和并Map输出文件，默认为True。
- hive.merge.mapredfiles = false：是否合并Reduce输出文件，默认为False。
- hive.merge.size.per.task = 256*1000*1000：合并文件的大小。

2. 小文件合并优化的例子

下面通过一个例子来综合说明如何针对小文件进行合并优化。

1）问题背景

企业Hive集群有张表，执行一次insert overwrite table select * from table语句大概需要7000s。源表是从某个数据库抽取过来的，用了500个Map Task。
该表以月为分区，每个分区文件夹下面产生了大量的小文件，有的都不到1MB。

2）问题描述

输入表本身就有很多小文件，插入的时候没有限制Reduce个数，也没有限制资源，导致产生很多个Reduce Task，进而产生多个小文件。

3）解决方案

（1）在Map输入的时候合并小文件，可设置的参数如下：

```
-- 每个Map最大输入大小，决定合并后的文件数
set mapred.max.split.size=256000000;
-- 一个节点上split的至少的大小，决定了多个DataNode上的文件是否需要合并
set mapred.min.split.size.per.node=100000000;
-- 一个交换机下split的至少的大小，决定了多个交换机上的文件是否需要合并
set mapred.min.split.size.per.rack=100000000;
-- 执行Map前进行小文件合并
set hive.input.format=org.apache.hadoop.hive.ql.io.CombineHiveInputFormat;
```

（2）解决方案：在Reduce输出的时候合并小文件，可设置的参数如下：

```
-- 在map-only job后合并文件，默认为true
set hive.merge.mapfiles = true;
-- 在map-reduce job后合并文件，默认为false
set hive.merge.mapredfiles = true;
```

```
-- 合并后每个文件的大小，默认为 256000000
set hive.merge.size.per.task = 256000000;
-- 平均文件大小，是决定是否执行合并操作的阈值，默认为 16000000
set hive.merge.smallfiles.avgsize = 100000000;
```

9.3.11　Group By

1. Map 端部分聚合

并不是所有的聚合操作都需要在 Reduce 端完成，很多聚合操作都可以先在 Map 端进行部分聚合，然后在 Reduce 端得出最终结果。

MapReduce 的 combiner 组件参数包括：

- hive.map.aggr = true：是否在 Map 端进行聚合，默认为 true。
- hive.groupby.mapaggr.checkinterval = 100000：在 Map 端进行聚合操作的条目数目。

2. 使用 Group By 出现数据倾斜的时候进行负载均衡

相关参数为 hive.groupby.skewindata = false。

当选项设定为 true 时，生成的查询计划会有两个 MapReduce job：第一个 MapReduce job 中，Map 的输出结果集合会随机分布到 Reduce 中，每个 Reduce 做部分聚合操作并输出结果，这样处理的结果是相同的 Group By Key 有可能被分发到不同的 Reduce 中，从而达到负载均衡的目的；第二个 MapReduce job 再根据预处理的数据结果按照 Group By Key 分布到 Reduce 中（这个过程可以保证相同的 Group By Key 被分布到同一个 Reduce 中），最后完成最终的聚合操作。

第10章

基于Hive的网站流量分析项目实战

本章基于电信运营商记录的用户手机上网访问某些网站行为的日志记录数据、手机相关的城市及运营商数据，进行数据清洗后形成数据仓库，然后基于 Hive 表进行流量统计和查询。没有 Java 编程背景的读者，在阅读过程中可以直接跳过 10.2 节，重点掌握利用 SQL 进行数据分析的方法。

主要内容包括：

- MapReduce 数据清洗。
- 构建 Hive 数据仓库表。
- 基于 Hive 表的查询。
- 基于 Hive 表的数据统计。

10.1　项目需求及分析

我们需要对用户手机上网访问某些网站行为的日志记录数据、手机相关的城市及运营商数据，进行数据清洗，形成数据仓库，然后基于 Hive 表进行流量统计和查询。

10.1.1　数据集及数据说明

这些数据包括两个文件。第一个文件是 http.log 日志文件，是电信运营商记录的用户手机上网访问某些网站行为的日志记录数据，其中上行流量+下行流量=总流量。

http.log 日志数据格式为（手机号码与网址之间是 Tab 键空格，其他各列之间是空格键空格）：

手机号码　　请求网站的 URL 上行流量（20 字节）下行流量（5000 字节）。

对应的实际数据示例为：

18611132889　　http://v.baidu.com/tv　20　5000。

部分日志数据内容如图 10-1 所示。再次提醒：手机号码和网址之间是个 Tab 键空格。

```
1563****688	http://v.baidu.com/movie 3936 12058
1390****439	http://movie.youku.com 10132 538
1519****948	https://image.baidu.com 19789 5238
1454****218	http://v.baidu.com/tv 7504 13253
1731****739	http://www.weibo.com/?category=7 7003 79
```

图 10-1　部分日志数据内容

第二个数据文件是 phone.txt，内容为手机号码段规则，是手机号码对应的城市和运营商的数据。phone.txt 数据格式为（各列之间都以 Tab 键空格隔开，比较整齐）：

手机号码前缀 手机号码段 手机号码对应的省份 城市 运营商 邮编 区号 行政划分代码

对应的实际数据示例为：

133　1332170　广西　南宁　电信　530000　0771　450100

部分数据内容如图 10-2 所示。

130	1300001	江苏	常州	联通	213000	0519	320400
130	1300002	安徽	巢湖	联通	238000	0551	340181
130	1300003	四川	宜宾	联通	644000	0831	511500
130	1300004	四川	自贡	联通	643000	0813	510300
130	1300005	陕西	西安	联通	710000	029	610100
130	1300006	江苏	南京	联通	210000	025	320100

图 10-2　部分数据内容

10.1.2　功能需求

数据分析的具体要求如下：

（1）根据用户上网日志记录数据，计算出总流量最高的网站 Top3，例如：v.baidu.com、weibo.com 等。

（2）根据用户上网日志记录数据，计算出总流量最高的手机号码 Top3。

（3）根据手机号码段归属地规则，计算出总流量最高的省份 Top3。

（4）根据手机号码段运营商规则，计算出总流量最高的运营商 Top2。

（5）根据手机号码段归属地规则，计算出总流量最高的城市 Top3。

10.2　利用 Java 实现数据清洗

本章讲解利用 Java 编程对相关数据进行清洗，因此需要搭建本地 Hadoop 运行环境，读者可参考 11.3.1 节内容。

本章用到的 Hadoop 框架采用伪分布式环境（实际上就是单个虚拟机），可以直接使用第 2 章搭建好的伪分布式环境，其服务器节点的 IP 地址为 192.168.56.201，服务器名称为 server201。在 VisualBox 中启动 CentOS7-201 虚拟机，并登录系统启动 Hadoop。

10.2.1 数据上传到 HDFS

首先需要将数据文件上传到 HDFS 中。具体操作是将数据文件 http.log、phone.txt 上传到 HDFS，保存的目录为/001/webdata/input，命令如下：

```
hdfs dfs -mkdir -p /001/webdata/input
hdfs dfs -put http.log /001/webdata/input
hdfs dfs -put phone.txt /001/webdata/input
```

10.2.2 http.log 数据清洗

编写 MapReduce 程序实现数据清洗，将不完整的数据过滤掉，保证输出数据以英文逗号（,）分隔，输出数据保存在两个文件中。

考虑项目后续要求等值连接，而 http.log 文件没有完全与 phone.txt 相同的字段，因此构造一个新字段，即手机号码的前 7 位，用这个字段连接 phone.txt 的手机号码段字段。其他具体数据处理的清洗逻辑参考代码注释。MapReduce 程序共 4 个类：用于传递多维度数据信息的 data1 类、用于数据预处理的 Mapper 类、用于数据预处理的 Reducer 类、可执行的主程序类。

（1）传递数据的 data1 类如代码 10.1 所示。

代码 10.1 data1.java

```
class data1 implements Writable{
   private String phonenum;//完整手机号码
   private String weburl;//截取的网站 URL
   private int liuliang;//总流量等于上行流量和下行流量之和
   private String phonenumpart;//手机号码前 7 位，方便后期多表查询

   public void write(DataOutput dataOutput) throws IOException {
      //序列化：将 java 对象转换为可跨机器传输数据流（二进制串/字节）的一种技术

      dataOutput.writeUTF(this.phonenum);
      dataOutput.writeUTF(this.weburl);
      dataOutput.writeInt(this.liuliang);
      dataOutput.writeUTF(this.phonenumpart);
   }

   public void readFields(DataInput dataInput) throws IOException {
      //反序列化：将可跨机器传输数据流（二进制串）转换为 java 对象的一种技术
      this.phonenum=dataInput.readUTF();
      this.weburl=dataInput.readUTF();
      this.liuliang=dataInput.readInt();
      this.phonenumpart=dataInput.readUTF();
   }
   public String getphonenum() {
      return phonenum;
   }
   public void setphonenum(String phonenum) {
      this.phonenum = phonenum;
   }
   public String getweburl() {
      return weburl;
```

```
        }
        public void setweburl(String weburl) {
            this.weburl = weburl;
        }
        public int getliuliang() {
            return liuliang;
        }
        public void setliuliang(int liuliang) {
            this.liuliang = liuliang;
        }
        public String getphonenumpart() {
            return phonenumpart;
        }
        public void setphonenumpart(String phonenumpart) {
            this.phonenumpart = phonenumpart;
        }
    }
```

（2）清除全部有缺失值的行，同时为了方便与 phone.txt 中的字段连接，截取手机号码前 7 位形成新字段。另外，需要生成总流量字段，便于后续统计。用于数据预处理的 Mapper 类如代码 10.2 所示。

代码 10.2　DMapper.java

```
    class DMapper extends Mapper<LongWritable,Text,NullWritable,data1>{
        protected void map(LongWritable key1, Text value1, Context context) throws
IOException, InterruptedException {
            String data=value1.toString();//获取一行数据
            String web="";
            String[] words=data.split("\t");//先通过"\t"把一行原数据分成长度为 2 的数组
            String[] words2=words[1].split(" ");//把 words[1]再次按照空格分成长度为 3
的数组
            if(!words[0].contains("1")){//因为 words[0]为手机号码，必包含数字 1，所以判断
是否包含数字 1，是则继续，否则退出方法
                return;
            }
            if (words2.length!=3){//因为 words 数组如果没有缺失值则数组长度必然等于 3，所以
判断长度是否等于 3，是则继续执行，否则退出方法
                return;
            }
            //用上述两个判断就可以清除全部有缺失值的行，一行中只要存在一个缺失值,即不会被传递到
reducer
            String part=words[0].substring(0,7);//截取手机号码前 7 位，方便后续多表连接
            data1 oo=new data1();//创建数据流量对象
            oo.setphonenumpart(part);//设置手机号码前 7 位
            int index1= words2[0].indexOf("/");//找到 URL 中第一个"/"的位置
            int index2= words2[0].indexOf("/",index1+5);//URL 中第三个"/"的位置
            if (index2==-1){//如果没有第三个"/"，则从第二个"/"之后一直截取到 URL 最后
                web= words2[0].substring(index1+2);
            }
            else{//如果有第三个"/"，则截取第二个"/"到第三个"/"中间的内容
                web= words2[0].substring(index1+2,index2);
            }

            oo.setweburl(web);//设置 URL
```

```
        oo.setphonenum(words[0]);//设置完整手机号码
        int sum=Integer.parseInt(words2[1])+Integer.parseInt(words2[2]);//计算
总流量
        oo.setliuliang(sum);//设置总流量
        context.write(NullWritable.get(),oo);//k2:NullWritable v2: 数据流量对象
    }
}
```

（3）用于数据预处理的 Reducer 类如代码 10.3 所示，得到的各个字段用英文逗号分隔。

代码 10.3 DReducer.java

```
class DReducer extends Reducer<NullWritable, data1,NullWritable,Text>{
    protected void reduce(NullWritable k3, Iterable<data1> v3, Context context)
throws IOException, InterruptedException {
        for (data1 oo:v3){
            //建立建表所需格式及数据
            String shuchu=oo.getphonenum()+","+oo.getphonenumpart()+","+
oo.getweburl()+","+oo.getliuliang();
            context.write(NullWritable.get(),new Text(shuchu));//k4:
NullWritable v4: TEXT 制表所需数据及格式
        }
    }
}
```

（4）用于 http.log 数据预处理的主程序类如代码 10.4 所示。

代码 10.4 webdata1_main.java

```
public class webdata1_main {
    public static void main(String[] args) throws Exception {
        //1. 创建一个 job 和任务入口（指定主类）
        Job job = Job.getInstance(new Configuration());
        job.setJarByClass(webdata1_main.class);

        //2. 指定 job 的 mapper 和输出的类型<k2 v2>
        job.setMapperClass(DMapper.class);
        job.setMapOutputKeyClass(NullWritable.class);
        job.setMapOutputValueClass(data1.class);

        //3. 指定 job 的 reducer 和输出的类型<k4  v4>
        job.setReducerClass(DReducer.class);
        job.setOutputKeyClass(NullWritable.class);
        job.setOutputValueClass(Text.class);

        //4. 指定 job 的输入和输出路径
        FileInputFormat.setInputPaths(job, new Path(args[0]));
        FileOutputFormat.setOutputPath(job, new Path(args[1]));
        //5. 执行 job
        job.waitForCompletion(true);
    }
}
```

程序打包：

```
mvn clean package
```

在运行 JAR 程序之前，需要定义除 CPU 和内存之外的新资源，因此检查 Hadoop 的配置目录/app/hadoop-3.2.3/conf/hadoop 下是否存在 resource-types.xml 文件，没有的话就创建一个，内容如下：

```
<?xml version="1.0" encoding="UTF-8"?>
<?xml-stylesheet type="text/xsl" href="configuration.xsl"?>

<configuration>
    <property>
        <name>yarn.resource-types</name>
        <value>resource1,resource2</value>
    </property>
    <property>
        <name>yarn.resource-types.resource1.units</name>
        <value>G</value>
    </property>
    <property>
        <name>yarn.resource-types.resource2.minimum-allocation</name>
        <value>1</value>
    </property>
    <property>
        <name>yarn.resource-types.resource2.maximum-allocation</name>
        <value>1024</value>
    </property>
</configuration>
```

把 JAR 文件（在配套下载资源中找）上传到/home/hadoop 目录，再运行程序：

```
hadoop jar httplog.jar webdata1_main /001/webdata/input/http.log
/001/webdata/output/01
```

使用如下命令查看结果：

```
hdfs dfs -cat /001/webdata/output/01/part-r-00000
```

输出结果如图 10-3 所示（输出数据已按“,”分隔）。

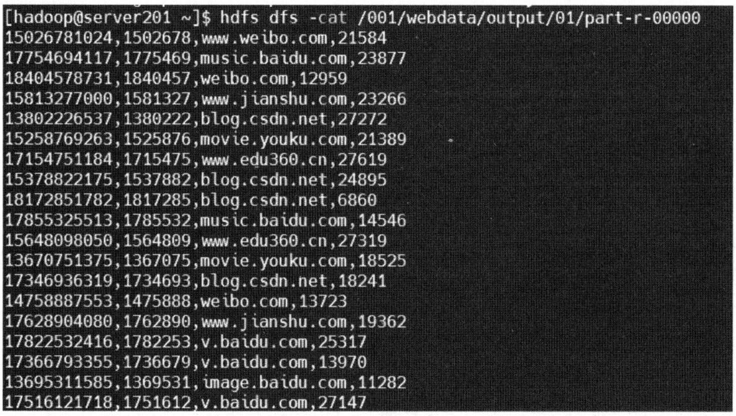

```
[hadoop@server201 ~]$ hdfs dfs -cat /001/webdata/output/01/part-r-00000
15026781024,1502678,www.weibo.com,21584
17754694117,1775469,music.baidu.com,23877
18404578731,1840457,weibo.com,12959
15813277000,1581327,www.jianshu.com,23266
13802226537,1380222,blog.csdn.net,27272
15258769263,1525876,movie.youku.com,21389
17154751184,1715475,www.edu360.cn,27619
15378822175,1537882,blog.csdn.net,24895
18172851782,1817285,blog.csdn.net,6860
17855325513,1785532,music.baidu.com,14546
15648098050,1564809,www.edu360.cn,27319
13670751375,1367075,movie.youku.com,18525
17346936319,1734693,blog.csdn.net,18241
14758887553,1475888,weibo.com,13723
17628904080,1762890,www.jianshu.com,19362
17822532416,1782253,v.baidu.com,25317
17366793355,1736679,v.baidu.com,13970
13695311585,1369531,image.baidu.com,11282
17516121718,1751612,v.baidu.com,27147
```

图 10-3　输出结果

10.2.3 phone.txt 数据清洗

编写用于处理 phone.txt 的第二个程序，数据处理的清洗逻辑参考代码注释。第二个程序共四个类：用于传递 phone 对象信息的 data1 类、用于数据预处理的 Mapper 类、用于数据预处理的 Reducer 类、可执行的主程序类。

（1）用于传递 phone 对象信息的 data1 类如代码 10.5 所示。

代码 10.5 data1.java

```
class data1 implements Writable{
    private String phonenumpart;//手机号码前 7 位，方便后期多表查询
    private String province;//省份
    private String yunyingshang;//运营商
    private String city;//城市

    public void write(DataOutput dataOutput) throws IOException {
        //序列化：将 java 对象转换为可跨机器传输数据流（二进制串/字节）的一种技术

        dataOutput.writeUTF(this.province);
        dataOutput.writeUTF(this.yunyingshang);
        dataOutput.writeUTF(this.city);
        dataOutput.writeUTF(this.phonenumpart);
    }

    public void readFields(DataInput dataInput) throws IOException {
        //反序列化：将可跨机器传输数据流（二进制串）转化为 java 对象的一种技术
        this.province=dataInput.readUTF();
        this.yunyingshang=dataInput.readUTF();
        this.city=dataInput.readUTF();
        this.phonenumpart=dataInput.readUTF();
    }
    public String getphonenumpart() {
        return phonenumpart;
    }
    public void setphonenumpart(String phonenumpart) {
        this.phonenumpart = phonenumpart;
    }
    public String getprovince() {
        return province;
    }
    public void setprovince(String province) {
        this.province = province;
    }
    public String getyunyingshang() {
        return yunyingshang;
    }
    public void setyunyingshang(String yunyingshang) {
        this.yunyingshang = yunyingshang;
    }
    public String getcity() {
        return city;
    }
    public void setcity(String city) {
```

```
        this.city = city;
    }
}
```

（2）清除全部有缺失值的行，同时创建手机信息对象。用于数据预处理的 Mapper 类如代码 10.6 所示。

代码 10.6　DMapper.java

```
class DMapper extends Mapper<LongWritable,Text,NullWritable,data1>{
    protected void map(LongWritable key1, Text value1, Context context) throws
IOException, InterruptedException {
        String data=value1.toString();//获取一行数据
        String[] words=data.split("\t");//通过“\t”把一行数据分成长度为 8 的数组
        if (   words[0].replace(" ","").equals("")||
               words[1].replace(" ","").equals("")||
               words[2].replace(" ","").equals("")||
               words[3].replace(" ","").equals("")||
               words[4].replace(" ","").equals("")||
               words[5].replace(" ","").equals("")||
               words[6].replace(" ","").equals("")||
               words[7].replace(" ","").equals("")){
            return;
        }//判断数组每个值是不是缺失值，如果是，则退出方法，不把这行数据传到 Reducer，如果
不是，则继续运行
        data1 oo=new data1();//创建数据流量对象
        oo.setphonenumpart(words[1]);//设置手机号码前 7 位
        oo.setprovince(words[2]);//设置省份
        oo.setyunyingshang(words[4]);//设置运营商
        oo.setcity(words[3]);//设置城市
        context.write(NullWritable.get(),oo);//k2:NullWritable v2：数据流量对象
    }
}
```

（3）用于数据预处理的 Reducer 类如代码 10.7 所示，得到的各个字段用英文逗号分隔。

代码 10.7　DReducer.java

```
class DReducer extends Reducer<NullWritable, data1,NullWritable,Text>{
    protected void reduce(NullWritable k3, Iterable<data1> v3, Context context)
throws IOException, InterruptedException {
        for (data1 oo:v3){
            //建立建表所需数据及对应格式
            String shuchu=oo.getphonenumpart()+","+oo.getprovince()+","+
oo.getyunyingshang()+","+oo.getcity();
            context.write(NullWritable.get(),new Text(shuchu));//k4:
NullWritable v4: TEXT 制表所需数据及格式
        }
    }
}
```

（4）用于 phone.txt 数据预处理的主程序类如代码 10.8 所示。

代码 10.8　webdata2_main.java

```
public class webdata2_main {
```

```
    public static void main(String[] args) throws Exception {
        //1. 创建一个 job 和任务入口（指定主类）
        Job job = Job.getInstance(new Configuration());
        job.setJarByClass(webdata2_main.class);

        //2. 指定 job 的 mapper 和输出的类型<k2 v2>
        job.setMapperClass(DMapper.class);
        job.setMapOutputKeyClass(NullWritable.class);
        job.setMapOutputValueClass(data1.class);

        //3. 指定 job 的 reducer 和输出的类型<k4  v4>
        job.setReducerClass(DReducer.class);
        job.setOutputKeyClass(NullWritable.class);
        job.setOutputValueClass(Text.class);

        //4. 指定 job 的输入和输出路径
        FileInputFormat.setInputPaths(job, new Path(args[0]));
        FileOutputFormat.setOutputPath(job, new Path(args[1]));
        //5. 执行 job
        job.waitForCompletion(true);
    }
}
```

程序打包：

```
mvn clean package
```

把 JAR 文件（在配套下载资源中找）上传到/home/hadoop 目录，再运行程序：

```
hadoop jar phonetxt.jar webdata2_main /001/webdata/input/phone.txt
/001/webdata/output/02
```

使用如下命令查看结果：

```
hdfs dfs -cat /001/webdata/output/02/part-r-00000
```

输出结果如图 10-4 所示，输出数据已按“,”分隔。

```
[hadoop@server201 ~]$ hdfs dfs -cat /001/webdata/output/02/part-r-00000
1999979,新疆,电信,石河了
1999978,新疆,电信,克孜勒苏
1999977,新疆,电信,阿克苏
1999976,新疆,电信,库尔勒
1999975,新疆,电信,阿克苏
1999974,新疆,电信,阿克苏
1999973,新疆,电信,阿克苏
1999972,新疆,电信,阿克苏
1999971,新疆,电信,阿克苏
1999970,新疆,电信,阿克苏
1999959,新疆,电信,石河子
1999958,新疆,电信,喀什
1999957,新疆,电信,博乐
1999956,新疆,电信,库尔勒
1999955,新疆,电信,喀什
1999954,新疆,电信,吐鲁番
1999953,新疆,电信,吐鲁番
```

图 10-4　输出结果

10.3　利用 MySQL 实现数据清洗

10.2.2 节使用 Java 编程实现了数据清洗。由于 http.log 日志文件和 phone.txt 手机号码段规则文件中的数据非常规整，因此我们可以利用 MySQL 实现数据清洗，从而在数据分析全过程中仅使用 SQL 语句完成本项目的数据分析工作。

同样，我们直接使用第 2 章搭建好的伪分布式环境，其服务器节点的 IP 地址为 192.168.56.201，服务器名称为 server201。在 VisualBox 中启动 CentOS7-201 虚拟机，并使用 hadoop 账号登录系统。

10.3.1　http.log 数据清洗

首先观察 http.log 数据特点，按空格可分为 3 部分，第 1 部分为“手机号码+Tab 键空格+url”，第 2 部分为上行流量，第 3 部分为下行流量。据此创建 3 个字段的表 httplog，把 http.log 文件数据导入表中，命令如下：

```
mysql> use weblog;
mysql> create table httplog(url varchar(200),shangxing int,xiaxing int);
mysql> load DATA infile '/tmp/http.log' into table httplog
character set utf8mb4 fields terminated by ' ' lines terminated by '\n';
```

查看一下表 httplog 中的数据，结果如图 10-5 所示。

```
mysql> select * from httplog limit 5;
```

```
mysql> select * from httplog limit 5;
+--------------------------------------------+-----------+---------+
| url                                        | shangxing | xiaxing |
+--------------------------------------------+-----------+---------+
| 15639120688   http://v.baidu.com/movie     |      3936 |   12058 |
| 13905256439   http://movie.youku.com       |     10132 |     538 |
| 15192566948   https://image.baidu.com      |     19789 |    5238 |
| 14542296218   http://v.baidu.com/tv        |      7504 |   13253 |
| 17314017739   http://www.weibo.com/?category=7 |   7003 |      79 |
+--------------------------------------------+-----------+---------+
5 rows in set (0.00 sec)
```

图 10-5　表 httplog 中的数据

由于第 1 部分数据中手机号码和网址没有分开，因此在表中加上手机号码字段，修改表的语句如下：

```
mysql> alter table httplog add phonenum char(11) first;
```

将手机号码从网址数据中取出并放入 phonenum 列中，结果如图 10-6 所示。

```
mysql> update httplog set  phonenum=SUBSTRING(url, 1,11 ) ;
```

```
mysql> update httplog set  phonenum=SUBSTRING(url, 1,11 ) ;
Query OK, 200000 rows affected (3.94 sec)
Rows matched: 200000  Changed: 200000  Warnings: 0

mysql> select * from httplog limit 5;
+-------------+------------------------------------------------------+-----------+---------+
| phonenum    | url                                                  | shangxing | xiaxing |
+-------------+------------------------------------------------------+-----------+---------+
| 15639120688 | 15639120688     http://v.baidu.com/movie             |      3936 |   12058 |
| 13905256439 | 13905256439     http://movie.youku.com               |     10132 |     538 |
| 15192566948 | 15192566948     https://image.baidu.com              |     19789 |    5238 |
| 14542296218 | 14542296218     http://v.baidu.com/tv                |      7504 |   13253 |
| 17314017739 | 17314017739     http://www.weibo.com/?category=7     |      7003 |      79 |
+-------------+------------------------------------------------------+-----------+---------+
5 rows in set (0.00 sec)
```

图 10-6 将手机号码从网址数据中取出并放入 phonenum 列中

删除 url 列中的手机号码+Tab 键空格，结果如图 10-7 所示。

```
mysql>  update httplog set  url=SUBSTRING(url, 13) ;
```

```
mysql> update httplog set  url=SUBSTRING(url, 13) ;
Query OK, 200000 rows affected (2.82 sec)
Rows matched: 200000  Changed: 200000  Warnings: 0

mysql> select * from httplog limit 5;
+-------------+----------------------------------+-----------+---------+
| phonenum    | url                              | shangxing | xiaxing |
+-------------+----------------------------------+-----------+---------+
| 15639120688 | http://v.baidu.com/movie         |      3936 |   12058 |
| 13905256439 | http://movie.youku.com           |     10132 |     538 |
| 15192566948 | https://image.baidu.com          |     19789 |    5238 |
| 14542296218 | http://v.baidu.com/tv            |      7504 |   13253 |
| 17314017739 | http://www.weibo.com/?category=7 |      7003 |      79 |
+-------------+----------------------------------+-----------+---------+
5 rows in set (0.01 sec)
```

图 10-7 删除 url 列中的手机号码+Tab 键空格

继续处理 url 字段，删除“//”及其前面的协议名，再删除网址中“/”及其后面的内容；把手机号码取前 7 位后放回去。再查询确认一下，结果如图 10-8 所示。

```
mysql>  update httplog set url=substring_index(url, '//', -1);
mysql>  update httplog set url=substring_index(url, '/', 1);
mysql>  update httplog set phonenum=substring(phonenum, 1, 7);
mysql>  select * from httplog limit 5;
```

```
mysql> select * from httplog limit 5;
+-------------+-----------------+-----------+---------+
| phonenum    | url             | shangxing | xiaxing |
+-------------+-----------------+-----------+---------+
| 15639120688 | v.baidu.com     |      3936 |   12058 |
| 13905256439 | movie.youku.com |     10132 |     538 |
| 15192566948 | image.baidu.com |     19789 |    5238 |
| 14542296218 | v.baidu.com     |      7504 |   13253 |
| 17314017739 | www.weibo.com   |      7003 |      79 |
+-------------+-----------------+-----------+---------+
5 rows in set (0.00 sec)
```

图 10-8 处理结果

接下来把表 httplog 导出为数据文件，并导入 Hive 数据库中备用：

```
mysql> select * from httplog into outfile "httplog.csv" fields terminated by
',' lines terminated by '\n';
```

导出来的数据文件到/var/lib/mysql/weblog/目录下查找。

10.3.2　phone.txt 数据清洗

phone.txt 文件的数据格式比较简单和整齐，按 Tab 键空格可分为 8 部分，即手机号码前缀、手机号码段、手机号码对应的省份、城市、运营商、邮编、区号、行政划分代码。据此创建 8 个字段的表 phonetxt，把 phone.txt 文件数据导入表 phonetxt 中，命令如下：

```
mysql> create table phonetxt(phoneno char(3),phonenumpart char(7),province
char(4),city char(10),yunyingshang char(8),zipcode char(20),quhao
char(4),xingzhengcode char(20));
mysql> load DATA infile '/tmp/phone.txt into table phonetxt character set utf8mb4
fields terminated by '\t' lines terminated by '\n';
```

查看一下表 phonetxt 中的数据，结果如图 10-9 所示。接下来就可以对数据做缺失值处理了，请读者自行完成。

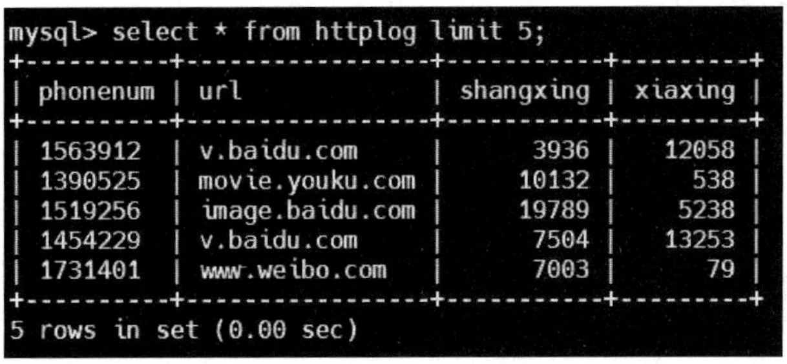

```
mysql> select * from httplog limit 5;
+----------+-----------------+-----------+---------+
| phonenum | url             | shangxing | xiaxing |
+----------+-----------------+-----------+---------+
| 1563912  | v.baidu.com     |      3936 |   12058 |
| 1390525  | movie.youku.com |     10132 |     538 |
| 1519256  | image.baidu.com |     19789 |    5238 |
| 1454229  | v.baidu.com     |      7504 |   13253 |
| 1731401  | www.weibo.com   |      7003 |      79 |
+----------+-----------------+-----------+---------+
5 rows in set (0.00 sec)
```

图 10-9　表 phonetxt 中的数据

本次数据分析只需要用到 4 个字段，即 phonenumpart、province、city、yunyingshang，接下来把表 phonetxt 中的这 4 个字段值导出为数据文件，并导入 Hive 数据库中备用：

```
mysql> select phonenumpart, province, city, yunyingshang from phonetxt into
outfile "phonetxt.csv" fields terminated by ',' lines terminated by '\n';
```

导出来的数据文件到/var/lib/mysql/weblog/目录下查找，如图 10-10 所示。

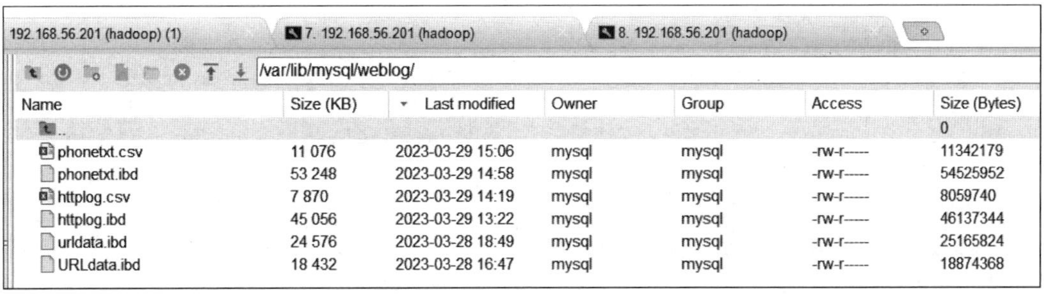

192.168.56.201 (hadoop) (1)　7. 192.168.56.201 (hadoop)　8. 192.168.56.201 (hadoop)

/var/lib/mysql/weblog/

Name	Size (KB)	Last modified	Owner	Group	Access	Size (Bytes)
..						0
phonetxt.csv	11 076	2023-03-29 15:06	mysql	mysql	-rw-r-----	11342179
phonetxt.ibd	53 248	2023-03-29 14:58	mysql	mysql	-rw-r-----	54525952
httplog.csv	7 870	2023-03-29 14:19	mysql	mysql	-rw-r-----	8059740
httplog.ibd	45 056	2023-03-29 13:22	mysql	mysql	-rw-r-----	46137344
urldata.ibd	24 576	2023-03-28 18:49	mysql	mysql	-rw-r-----	25165824
URLdata.ibd	18 432	2023-03-28 16:47	mysql	mysql	-rw-r-----	18874368

图 10-10　导出来的数据文件所在的目录

10.4 数据分析的实现

10.4.1 创建 Hive 库和表

（1）创建 Hive 库：

```
create database if not exists hwy001;
```

（2）使用自己的数据库：

```
use hwy001;
```

（3）创建第一张表 URLdata，用于保存 http.log 经程序处理后的结果数据：

```
create table URLdata(phonenum string,phonenumpart string,web string,money int)
row format delimited fields terminated by ',';
```

（4）创建第二张表 phoneInfo，用于保存 phone.txt 经程序处理后的结果数据：

```
create table phoneInfo(phonenumpart string,province string,yunyingshang
string,city string) row format delimited fields terminated by ',';
```

（5）将清洗后的数据导入 Hive，向两张表中导入数据：

```
load data inpath '/001/webdata/output/01/part-r-00000' into table URLdata;
load data inpath '/001/webdata/output/02/part-r-00000' into table phoneInfo;
```

10.4.2 使用 SQL 进行数据分析

1. 根据用户上网日志记录数据，计算出总流量最高的网站 Top3（例如：v.baidu.com、weibo.com）

命令如下：

```
select web,sum(money) as mm from URLdata group by web order by mm desc limit
3;
```

运行过程的提示信息如下：

```
hive> load data inpath '/001/webdata/output/02/part-r-00000' into table
phoneInfo;
Loading data to table hwy001.phoneinfo
OK
Time taken: 0.381 seconds
hive> select web,sum(money) as mm from URLdata group by web order by mm desc
limit 3;
Query ID = hadoop_20230319164359_4221c9de-ae3d-4f94-ac57-1e8bbe53faec
Total jobs = 2
Launching Job 1 out of 2
Number of reduce tasks not specified. Estimated from input data size: 1
In order to change the average load for a reducer (in bytes):
  set hive.exec.reducers.bytes.per.reducer=<number>
In order to limit the maximum number of reducers:
  set hive.exec.reducers.max=<number>
```

```
    In order to set a constant number of reducers:
      set mapreduce.job.reduces=<number>
    Starting Job = job_1679213490837_0004, Tracking URL =
http://server201:8088/proxy/application_1679213490837_0004/
    Kill Command = /app/hadoop-3.2.3/bin/mapred job  -kill job_1679213490837_0004
    Hadoop job information for Stage-1: number of mappers: 1; number of reducers: 1
    2023-03-19 16:44:11,446 Stage-1 map = 0%,  reduce = 0%
    2023-03-19 16:44:18,853 Stage-1 map = 100%,  reduce = 0%, Cumulative CPU 1.89 sec
    2023-03-19 16:44:24,153 Stage-1 map = 100%,  reduce = 100%, Cumulative CPU 2.9
sec
    MapReduce Total cumulative CPU time: 2 seconds 900 msec
    Ended Job = job_1679213490837_0004
    Launching Job 2 out of 2
    Number of reduce tasks determined at compile time: 1
    In order to change the average load for a reducer (in bytes):
      set hive.exec.reducers.bytes.per.reducer=<number>
    In order to limit the maximum number of reducers:
      set hive.exec.reducers.max=<number>
    In order to set a constant number of reducers:
      set mapreduce.job.reduces=<number>
    Starting Job = job_1679213490837_0005, Tracking URL =
http://server201:8088/proxy/application_1679213490837_0005/
    Kill Command = /app/hadoop-3.2.3/bin/mapred job  -kill job_1679213490837_0005
    Hadoop job information for Stage-2: number of mappers: 1; number of reducers: 1
    2023-03-19 16:44:37,543 Stage-2 map = 0%,  reduce = 0%
    2023-03-19 16:44:43,860 Stage-2 map = 100%,  reduce = 0%, Cumulative CPU 0.96 sec
    2023-03-19 16:44:50,097 Stage-2 map = 100%,  reduce = 100%, Cumulative CPU 2.03 sec
    MapReduce Total cumulative CPU time: 2 seconds 30 msec
    Ended Job = job_1679213490837_0005
    MapReduce Jobs Launched:
    Stage-Stage-1: Map: 1  Reduce: 1   Cumulative CPU: 2.9 sec   HDFS Read: 8069170
HDFS Write: 461 SUCCESS
    Stage-Stage-2: Map: 1  Reduce: 1   Cumulative CPU: 2.03 sec   HDFS Read: 8099 HDFS
Write: 195 SUCCESS
    Total MapReduce CPU Time Spent: 4 seconds 930 msec
    OK
    www.jianshu.com 717979222
    v.baidu.com     678201065
    www.edu360.cn   482997506
    Time taken: 51.196 seconds, Fetched: 3 row(s)
    hive>
```

最终的运行结果如图 10-11 所示。

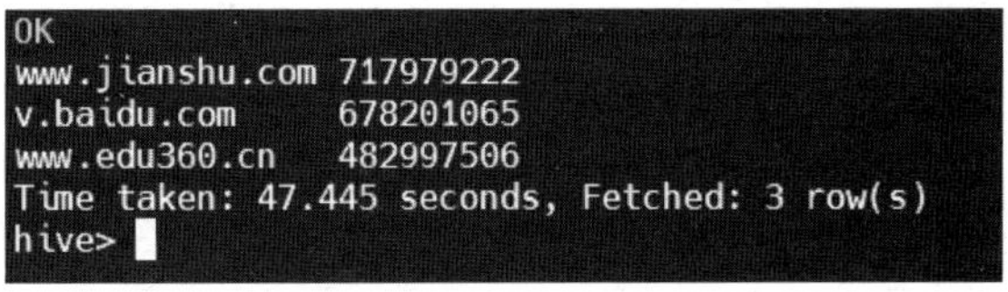

```
OK
www.jianshu.com 717979222
v.baidu.com     678201065
www.edu360.cn   482997506
Time taken: 47.445 seconds, Fetched: 3 row(s)
hive>
```

图 10-11　运行结果

从运行结果可以看出，Hive 执行 SQL 语句实际上就是执行 MapReduce 任务。

2. 根据用户上网日志记录数据，计算出总流量最高的手机号码 Top3

命令如下：

```
select phonenum,sum(money) as mm from URLdata group by phonenum order by mm desc limit 3;
```

运行过程的提示信息略去。运行结果如图 10-12 所示。

```
OK
17859739080     43735
15972344746     39989
15680893723     39943
Time taken: 56.205 seconds, Fetched: 3 row(s)
hive>
```

图 10-12　运行结果

3. 根据手机号码段归属地规则，计算出总流量最高的省份 Top3

命令如下：

```
select province,sum(money) as s from phoneInfo as p join URLdata as u on p.phonenumpart=u.phonenumpart group by province ORDER BY s DESC limit 3;
```

运行过程的提示信息略去。运行结果如图 10-13 所示。

```
OK
广东    438816394
江苏    279100752
山东    274846877
Time taken: 58.652 seconds, Fetched: 3 row(s)
hive>
```

图 10-13　运行结果

4. 根据手机号码段运营商规则，计算出总流量最高的运营商 Top2

命令如下：

```
select yunyingshang,sum(money) s from phoneInfo as p join URLdata as u on p.phonenumpart=u.phonenumpart group by yunyingshang ORDER BY s DESC limit 2;
```

运行过程的提示信息略去。运行结果如图 10-14 所示。

```
OK
移动    1977129164
联通    1039736761
Time taken: 57.46 seconds, Fetched: 2 row(s)
hive>
```

图 10-14　运行结果

5. 根据手机号码段归属地规则，计算出总流量最高的城市 Top3

命令如下：

```
select city,sum(money) as s from phoneInfo as p join URLdata as u on p.phonenumpart=u.phonenumpart group by city ORDER BY s DESC limit 3;
```

运行结果如图 10-15 所示。

```
OK
北京      160146308
上海      125214342
广州      103100601
Time taken: 57.123 seconds, Fetched: 3 row(s)
hive>
```

图 10-15　运行结果

第11章 旅游酒店评价大数据分析项目实战

本章将通过一个城市旅游酒店的基本数据及用户评论数据集合，综合运用 Hadoop 进行大数据分析项目实战。项目基于山东省青岛市酒店基本数据和用户评论数据，使用 HDFS 应用程序进行上传和存储，然后基于 HDFS 运用 MapReduce 进行数据预处理，处理之后的数据构建 Hive 数据仓库，并结合用户需求生成数据仓库分析结果表，将表导出到 MySQL，最后构建基于 Spring Boot 框架的 Web 项目，并结合 ECharts 技术实现数据可视化呈现，为企业和用户提供决策支持。没有 Java 编程背景的读者，在阅读过程中可以直接跳过 11.3 节、11.6 节内容，重点掌握利用 SQL 进行数据分析的方法。

主要内容包括：

- Hadoop HDFS 数据存储。
- 构建 Hive 数据仓库表。
- 基于 Hadoop MapReduce 或 SQL 的数据清洗。
- Hive 数据导出到 MySQL。
- 基于 ECharts 的可视化展示。

11.1 项 目 介 绍

随着计算机网络的发展，各大型网站及平台实时更新，产生了大量的数据。在当今大数据背景下，各行各业积累了海量数据，这些数据具有数据容量大、类型多、数据增长速度快、价值密度高的特点。许多学者也展开了关于大数据分析算法、分析模式及分析软件工具方面的研究，其中，在大数据结构模型和数据科学理论体系、大数据分析和挖掘基础理论方面有很大的进步，大数据的应用领域也从科学、工程、电信等领域扩展到各行各业。

在我国，许多规模较大的酒店都有自己的酒店管理系统，提供了完善的酒店管理和酒店预订、评价等服务。部分中小型酒店由于缺乏资金，依托第三方平台提供在线服务，客户进行操作后，第三方平台会生成记录保存下来。酒店长期积累了大量的在线基本数据和用户评论数据。针对酒店行业，如何利用大数据技术来对现有的数据进行处理和分析，为酒店从业者和出行用户提供直观的参考决策，是当前急需解决的问题。一方面，根据用户在线评论数据，帮助酒店从业者提供直观的决策支持，改善酒店管理，以获取最大利润；另一方面，提供某地区的酒店基本满意度情况、酒店分布情况、热门酒店等可视化图表，为用户出行提供可靠参考。

为了使用户对旅游目标城市的酒店住宿和用户满意度、城市各地区酒店分布、用户出游目的等情况有更加直观、明确的了解，并为用户提前规划好住宿和旅游景点的选择提供决策支持，本项目基于山东省青岛市酒店基本数据和用户评论数据构建大数据平台数据仓库，并进行统计分析，最后以 Web 网页的形式将分析结果以可视化图表方式进行展示，也为酒店从业者提供一定的决策支持，方便他们在前期市场调查过程中提前了解各区酒店分布、满意度、用户出游目的等信息。比如，可以根据用户出游类型占比等信息来为酒店从业者规划酒店类型及相关配套。

本项目具体过程为：对青岛市的酒店评论和酒店基本数据进行大数据分析和处理，将数据存储到 Hadoop 集群，经过数据清洗后构建 Hive 数据仓库，并基于 Hive 仓库进行数据分析，将分析结果导入 MySQL 中，最后构建基于 Java EE 的 Web 项目进行酒店数据可视化展示。

本章项目运行前需要搭建 Hadoop 基础环境，这里为了方便展开项目，采用了伪分布式的 Hadoop 集群，服务器的 IP 地址为 192.168.56.201，采用 CentOS 7 操作系统，Hive、MySQL 等相关软件安装在该服务器上。开发环境在 Windows 本地主机上，使用 IDEA 作为开发工具。

11.2　项目需求及分析

11.2.1　数据集及数据说明

为了给游客提供城市酒店满意度、酒店分布等出行信息，项目对数据集有一定的要求。本项目已提供山东省青岛市的酒店基本数据和酒店评论数据两个数据集文件。下面对这两个文件的属性进行说明。

首先是用户评论数据，文件名为 hoteldata.csv。对于酒店用户评论数据采集的数据格式，描述为：

- 酒店评论的用户名（user_name），便于对评论进行用户画像。
- 酒店名字（hotel_name），便于统计酒店数量。
- 出游类型（trip_type），表示用户旅游的性质。
- 用户评价时间（evaluate_time）。
- 酒店评论的发布日期（post_time），便于了解是否是最近发布的。
- 用户对酒店的评分（user_score），这个数据属性能和用户评论进行映射，便于后面的情感分析训练集的制作。

其次是酒店基本数据，文件名为 hotelbasic.csv。对于酒店基本数据采集的数据格式，描述为：

- 酒店 id。
- 酒店名称。

- 酒店评分。
- 酒店地址。
- 酒店星级。
- 星级详情。

以下是两个数据文件中的部分内容：

（1）部分酒店基本数据属性及原始数据如下：

```
60169364,枫叶酒店式公寓（青岛金沙滩传媒广场店）,4.7,黄岛区珠江路 588 号传媒广场天相公寓 4 号楼北侧 1 号 101 室。 （ 黄岛金沙滩度假区）,hotel_diamond02,经济型
6841087,青岛新天桥快捷宾馆,4.4,市南区肥城路 51-1 号。 （ 青岛火车站/栈桥/中山路劈柴院）,hotel_diamond02,经济型
4640468,欧圣兰廷公寓(青岛万达东方影都店),4.5,黄岛区滨海大道万达公馆 A1 区 2 号楼办理入住。( 西海岸度假区）,hotel_diamond02,经济型
4539535,世纪双帆海景度假酒店(青岛城市阳台店),4.4,黄岛区滨海大道1288 号那鲁湾 1 号楼。( 西海岸度假区）,hotel_diamond02,经济型
60664769,欧圣兰廷度假公寓（青岛城市阳台店）,4.7,黄岛区滨海大道 4098 号城市阳台风景区世茂悦海 13 号楼一楼。 （ 西海岸度假区）,hotel_diamond02,经济型
823437,青岛花园大酒店—贵宾楼,4.8,市南区彰化路 6 号贵宾楼。 （ 五四广场/万象城/奥帆中心/市政府青岛大学）,hotel_diamond04,高档型
43679094,青岛燕岛之星度假公寓,4.7,市南区燕儿岛路 15 号 1 楼大厅。 （ 五四广场/万象城/奥帆中心/市政府）,hotel_diamond02,经济型
22416789,容锦酒店（青岛台东步行街店）,4.6,市北区延安路 129 号利群百惠商厦 6-7 楼。 （ 台东步行街/啤酒街）,hotel_diamond02,经济型
17263343,慢居听海酒店(青岛吾悦广场店),4.8,黄岛区滨海大道 2888 号梦时代广场 17 号楼 1 楼大厅 101 室。 （ 黄岛金沙滩度假区）,hotel_diamond02,经济型
427962,安澜宾舍酒店（青岛东海中路海滨店）,4.4,市南区东海中路 30 号银海大世界院内。 （ 五四广场/万象城/奥帆中心/市政府）,hotel_diamond02,经济型
21122269,青岛蓝朵海景假日公寓,4.8,崂山区秦岭路 19 号协信中心 3 号楼 28 层。 ( 国际会展中心/石老人海水浴场）,,经济型
```

（2）部分酒店用户评论属性及原始数据如下（评论内容列已删掉）：

```
Elainemimi,枫叶酒店式公寓（青岛金沙滩传媒广场店）,其他,20-Jul,2020/7/6,5
WeChat268192****,枫叶酒店式公寓（青岛金沙滩传媒广场店）,商务出差,20-Jun,2020/6/28,5
CFT010000002415****,枫叶酒店式公寓（青岛金沙滩传媒广场店）,朋友出游,20-Jun,2020/6/26,5
WeChat320342****,枫叶酒店式公寓（青岛金沙滩传媒广场店）,情侣出游,20-Jun,2020/6/20,5
WeChat320342****,枫叶酒店式公寓（青岛金沙滩传媒广场店）,情侣出游,20-Jun,2020/6/21,5
WeChat320342****,枫叶酒店式公寓（青岛金沙滩传媒广场店）,情侣出游,20-Jun,2020/6/21,5
M24589****,枫叶酒店式公寓（青岛金沙滩传媒广场店）,独自旅行,20-May,2020/6/4,5
M221656****,青岛新天桥快捷宾馆,家庭亲子,20-Aug,2020/8/27,5
M419190****,青岛新天桥快捷宾馆,家庭亲子,20-Jul,2020/8/10,5
M354977****,青岛新天桥快捷宾馆,独自旅行,20-Jul,2020/7/30,5
M355264****,青岛新天桥快捷宾馆,独自旅行,20-Jul,2020/8/3,5
M311789****,青岛新天桥快捷宾馆,其他,20-Jul,2020/8/3,5
朝生牧者,青岛新天桥快捷宾馆,独自旅行,20-Jul,2020/8/10,5
WeChat229580****,青岛新天桥快捷宾馆,情侣出游,20-Aug,2020/8/26,4.5
M415985****,青岛新天桥快捷宾馆,独自旅行,20-Jul,2020/8/6,5
WeChat385461****,青岛新天桥快捷宾馆,朋友出游,20-Aug,2020/8/7,3.8
M416989****,青岛新天桥快捷宾馆,商务出差,20-Jul,2020/8/10,5
```

```
M272306****,青岛新天桥快捷宾馆,情侣出游,20-Jun,2020/6/22,5
w 风清,青岛新天桥快捷宾馆,情侣出游,20-Jun,2020/8/10,5
WeChat290262****,青岛新天桥快捷宾馆,朋友出游,20-Jun,2020/8/10,5
杰科 1105,青岛新天桥快捷宾馆,家庭亲子,20-Jul,2020/7/22,5
M317651****,青岛新天桥快捷宾馆,家庭亲子,20-Aug,2020/8/18,3.8
通帕蓬廖化,青岛新天桥快捷宾馆,独自旅行,20-Aug,2020/8/19,4.8
M327775****,青岛新天桥快捷宾馆,朋友出游,20-Aug,2020/8/11,5
WeChat270171****,青岛新天桥快捷宾馆,家庭亲子,20-Aug,2020/8/6,5
平哥儿走天涯,青岛新天桥快捷宾馆,商务出差,20-Jul,2020/8/7,5
WeChat266127****,青岛新天桥快捷宾馆,朋友出游,20-Aug,2020/8/12,5
```

数据清洗的主要工作说明如下：

（1）酒店基本数据集中的酒店星级类型这一列数据叫法不一致，比如有的酒店叫四星级，有的则叫高档型，这里统一做一下处理，将所有的“国家旅游局评定为五星级”替换为“豪华型”，将“国家旅游局评定为四星级”替换为“高档型”，将“国家旅游局评定为三星级”替换为“舒适型”，将“国家旅游局评定为二星级”替换为“经济型”。

（2）由于大数据服务平台这个子模块并不对用户具体评论内容进行情感分析，情感分析是交给情感分析子系统来处理的，因此这里将评论内容数据列删掉。给出的数据文件中已经删掉了用户评论。

（3）删除所有空行。

（4）从酒店地址中提取出区县名称替换掉地址那一列内容，为区县酒店分布统计提供标准数据。

11.2.2　功能需求

本项目对山东省青岛市的酒店数据进行大数据分析和处理。首先需要采集青岛市的酒店基本信息数据和用户评论数据，之后将采集到的数据集上传到 Hadoop 平台的 HDFS 来存储，然后基于 Hadoop 进行数据清洗，以及基于 HDFS 中的两个数据集构建 Hive 数据仓库，最后基于 Hive 数据仓库，从用户关心的酒店及评论信息的维度进行数据分析处理。

本项目系统的主要数据分析流程如图 11-1 所示。

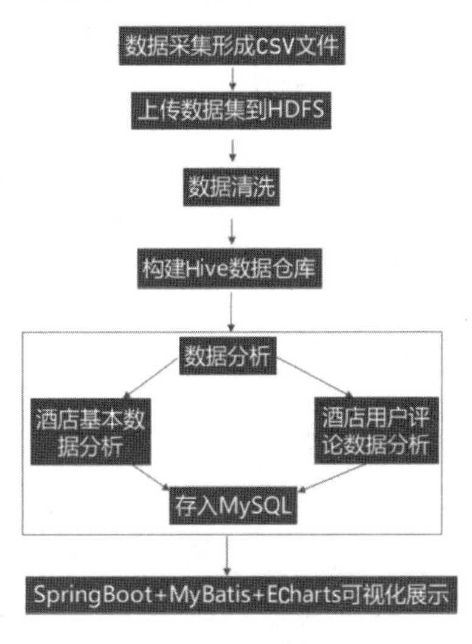

图 11-1　数据分析流程

根据系统功能要求分为 5 个关注角度，并为每一个关注角度创建 Hive 内部表，具体如下：

（1）用户印象统计，也是用户对该地区的总体满意度情况。用户对住过的酒店发表评论，同时也可以打分。以下是根据用户对该地区或城市的酒店的总体评分情况来统计用户的总体印象：4.5～5 分为优良、3.5～4.5 为良好、3.5 分以下为差，用于统计酒店用户评分的等级比例。

（2）统计在线评论数最多的十大酒店，即十大网络人气酒店。一般情况下，一家酒店的评论数量能代表这家酒店的人气，这里统计的是酒店名称和评论数目。

（3）不同旅游类型占比统计。根据用户评论 Hive 外部表 hotel_data 进行不同旅游类型的统计。根据旅游类型与用户满意度情况，为用户旅游出行提供参考，了解该地区更适合哪种类型的旅游。

（4）酒店星级分布情况统计。设计酒店星级和数量两个属性，显示不同星级的酒店数量占比，为不同层次的用户提供星级酒店的数量分布。

（5）城市不同地区的酒店数量分布情况。以热力图方式呈现，同时需要显示每个地区的酒店数量和平均评论得分情况。

最后，利用 Java 编程将产生的 Hive 内部表数据导出到 MySQL 数据库。

数据可视化部分，根据用户的 5 个关注角度的分析结果，构建用于数据展现的 Web 项目，采用的技术是 Spring Boot+MyBatis+MySQL，开发工具是 IDEA，图表采用 Echarts 来提供页面图表渲染支持。为了提升页面加载速度和用户体验，采用 Ajax 异步加载的方式来进行图表呈现。

11.3 利用 Java 实现数据清洗

本项目运行环境采用 Hadoop 伪分布式环境（实际上就是单个虚拟机），可以直接使用第 2 章搭建好的伪分布式环境，其虚拟机名称为 CentOS7-201，服务器节点的 IP 地址为 192.168.56.201，服务器名称为 server201。

11.3.1 本地 Hadoop 运行环境搭建

由于本项目主要使用本地 Java 程序与远程 Hive 服务器交互，实现数据文件的上传、数据清洗、数据导入 MySQL 以及数据可视化等操作，因此我们需要在本地搭建 Hadoop 运行环境，以方便编写的代码引用 Hadoop 的 JAR 包。

步骤 01 首先把 hadoop-3.2.3 软件包在本地解压缩，比如笔者本地目录为 c:\x\hadoop-3.2.3，目录结构如图 11-2 所示。

› 此电脑 › OS (C:) › x › hadoop-3.2.3

名称	修改日期	类型	大小
bin	3/18/2023 9:46 AM	文件夹	
etc	3/18/2023 9:46 AM	文件夹	
include	3/18/2023 9:46 AM	文件夹	
lib	3/18/2023 9:46 AM	文件夹	
libexec	3/18/2023 9:46 AM	文件夹	
sbin	3/18/2023 9:46 AM	文件夹	
share	3/18/2023 9:46 AM	文件夹	
LICENSE.txt	3/10/2022 1:39 PM	Text Document	148 KB
NOTICE.txt	3/10/2022 1:39 PM	Text Document	22 KB

图 11-2　hadoop-3.2.3 目录结构

步骤 02 下载 winutils.exe 文件。

winutils.exe 是在 Window 系统上运行 Hadoop 所需要的，因为 Hadoop 主要基于 Linux 编写，这个 winutil.exe 可以模拟 Linux 下的目录环境，所以 Hadoop 放在 Windows 下运行的时候就需要这个辅助程序。winutils.exe 下载地址为：

```
https://github.com/steveloughran/winutils
```

下载页面如图 11-3 所示。下载下来的文件名为 winutils-master.zip，解压这个文件，把 hadoop-3.0.0\bin 目录下的文件全部复制到本地 hadoop-3.2.3 的 bin 目录下，比如 c:\x\hadoop-3.2.3\bin。

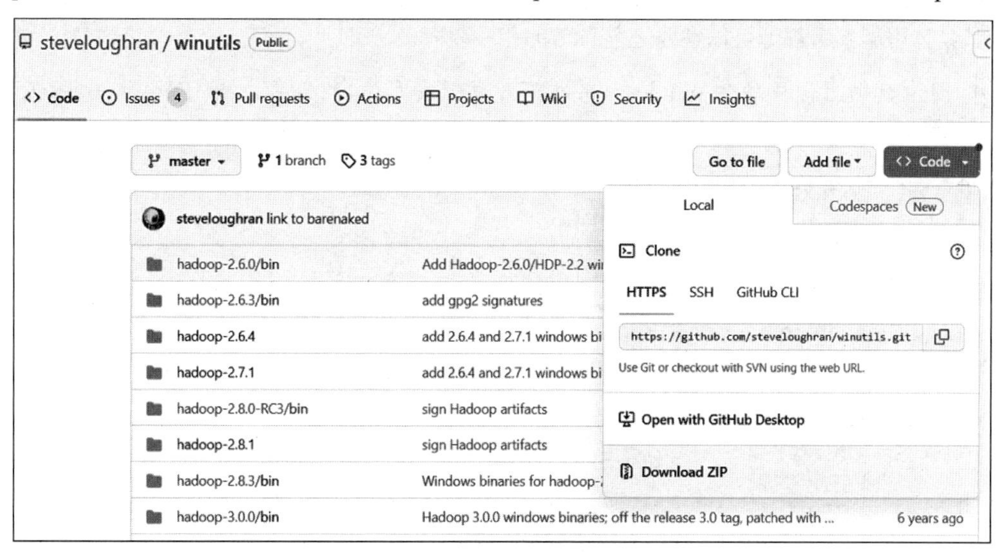

图 11-3　下载 winutils.exe

步骤 03 在本地配置环境变量。

首先根据 Java 安装路径配置 JAVA_HOME，然后配置 HADOOP_HOME 为 C:\x\hadoop-3.2.3，在 PATH 中加入%HADOOP_HOME%\bin，再把 C:\x\hadoop-3.2.3\share\hadoop 下所有包含 JAR 包

的目录加入 CLASS_PATH 中，这些目录如图 11-4 所示。在本地运行数据上传、数据清洗等程序时需要用到相关 JAR 包。

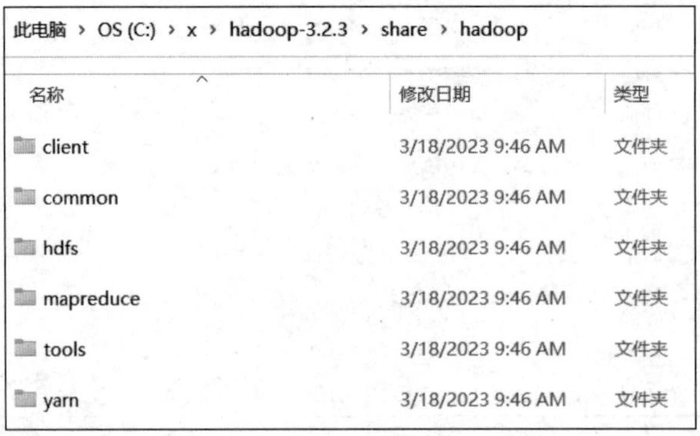

此电脑 › OS (C:) › x › hadoop-3.2.3 › share › hadoop

名称	修改日期	类型
client	3/18/2023 9:46 AM	文件夹
common	3/18/2023 9:46 AM	文件夹
hdfs	3/18/2023 9:46 AM	文件夹
mapreduce	3/18/2023 9:46 AM	文件夹
tools	3/18/2023 9:46 AM	文件夹
yarn	3/18/2023 9:46 AM	文件夹

图 11-4 hadoop-3.2.3\share\hadoop 下的目录

步骤 04 将 winutil.exe 所在目录中的 hadoop.dll 文件复制一份，放到本地 windows/system32 下面，然后重新启动计算机（或者重新启动 IDEA）。

至此，本地 Hadoop 运行环境搭建完毕。

11.3.2 数据上传到 HDFS

将经过数据预处理后的酒店数据集以及用户评论数据集对应的 CSV 文件上传到 HDFS 存储。这里为了方便调试代码，将程序部分放到本地 Windows 系统上开发和运行，将 Hadoop 安装到 CentOS 7 系统，具体框架安装部分可以参照之前章节的相关内容。

步骤 01 因为数据上传程序在 Windows 系统下开发开发和运行，所以打开 IDEA 新建一个 Maven 项目 pom.xml，用于数据上传和数据清洗。项目 pom.xml 的代码如下：

代码 11.1 pom.xml

```
<!--设置依赖版本号-->
    <properties>
        <scala.version>2.11.8</scala.version>
        <hadoop.version>3.2.3</hadoop.version>
        <spark.version>3.1.1</spark.version>
    </properties>
    <dependencies>
        <!--Scala-->
        <dependency>
            <groupId>org.scala-lang</groupId>
            <artifactId>scala-library</artifactId>
            <version>${scala.version}</version>
        </dependency>
        <!--Spark-->
        <dependency>
            <groupId>org.apache.spark</groupId>
            <artifactId>spark-core_2.11</artifactId>
            <version>${spark.version}</version>
        </dependency>
```

```
        <!--Hadoop-->
        <dependency>
            <groupId>org.apache.hadoop</groupId>
            <artifactId>hadoop-client</artifactId>
            <version>${hadoop.version}</version>
        </dependency>
        <dependency>
            <groupId>org.apache.hadoop</groupId>
            <artifactId>hadoop-common</artifactId>
            <version>${hadoop.version}</version>
        </dependency>
        <dependency>
            <groupId>org.apache.hadoop</groupId>
            <artifactId>hadoop-hdfs</artifactId>
            <version>${hadoop.version}</version>
        </dependency>
        <dependency>
            <groupId>org.apache.hadoop</groupId>
            <artifactId>hadoop-mapreduce-client-core</artifactId>
            <version>${hadoop.version}</version>
        </dependency>
        <dependency>
            <groupId>junit</groupId>
            <artifactId>junit</artifactId>
            <version>4.12</version>
        </dependency>
        <dependency>
            <groupId>org.apache.zookeeper</groupId>
            <artifactId>zookeeper</artifactId>
            <version>3.4.10</version>
        </dependency>
        <!--  https://mvnrepository.com/artifact/com.google.code.gson/gson
        <dependency>
            <groupId>com.google.code.gson</groupId>
            <artifactId>gson</artifactId>
            <version>2.8.0</version>
        </dependency>
        &lt;!&ndash;
https://mvnrepository.com/artifact/org.apache.kafka/kafka &ndash;&gt;
        <dependency>
            <groupId>org.apache.kafka</groupId>
            <artifactId>kafka_2.11</artifactId>
            <version>1.0.0</version>
        </dependency>-->
    </dependencies>
```

步骤 02 编写具体上传文件的 Java 程序，代码如下：

代码 11.2　HDFS_CRUD.java

```
import java.io.FileNotFoundException;
import java.io.IOException;
import org.apache.hadoop.conf.Configuration;
import org.apache.hadoop.fs.BlockLocation;
import org.apache.hadoop.fs.FileStatus;
```

```
import org.apache.hadoop.fs.FileSystem;
import org.apache.hadoop.fs.LocatedFileStatus;
import org.apache.hadoop.fs.Path;
import org.apache.hadoop.fs.RemoteIterator;
import org.junit.Before;
import org.junit.Test;
public class HDFS_CRUD {
    FileSystem fs = null;
    @Before
    public void init() throws Exception {
        // 构造一个配置参数对象，设置一个参数：我们要访问的 HDFS 的 URI
        Configuration conf = new Configuration();
        // 这里指定使用的是 HDFS 文件系统
        conf.set("fs.defaultFS", "hdfs://server201:8020");
        // 这个解决 HDFS 问题
        conf.set("fs.hdfs.impl",
org.apache.hadoop.hdfs.DistributedFileSystem.class.getName());
        // 这个解决本地 file 问题
        conf.set("fs.file.impl",
org.apache.hadoop.fs.LocalFileSystem.class.getName());
        // 通过如下的方式进行客户端身份的设置
        System.setProperty("HADOOP_USER_NAME", "root");
        System.setProperty("HADOOP_HOME", "/app/hdoop-3.2.3");
        // 通过 FileSystem 的静态方法获取文件系统客户端对象
        fs = FileSystem.get(conf);
    }
    /**
     * 将本地的爬取的酒店数据和评论原始数据上传到 HDFS
     * @throws IOException
     */
    @Test
    public void testAddFileToHdfs() throws IOException {
        // 要上传的文件所在的本地路径
        Path src = new Path("D:/data/hoteldata.csv");
        // 要上传到 HDFS 的目标路径
        Path dst = new Path("/hdfsdata");
        // 上传文件方法
        fs.copyFromLocalFile(src, dst);
         src = new Path("D:/data/hotelbasic.csv");
        // 要上传到 HDFS 的目标路径
         dst = new Path("/hdfsbasic");
        // 上传文件方法
        fs.copyFromLocalFile(src, dst);
        // 关闭资源
        fs.close();
    }
}
```

项目中已经包含了数据集文件夹，里面有两个文件，其中，hotelbasic.csv 是酒店基本数据集，hoteldata.csv 是酒店评论数据集。通过以上程序，把这两个文件上传到 HDFS 中。

上传数据集后，在 Hadoop 平台使用 hadoop fs -ls 命令查看，可以看到多出两个路径/hdfsbasic 和/hdfsdata，里边的文件分别对应基本数据和评论数据，如图 11-5 所示。

```
[hadoop@server201 ~]$ hdfs dfs -ls /
Found 7 items
drwxr-xr-x   - root   supergroup          0 2023-03-21 13:41 /a
drwxr-xr-x   - root   supergroup          0 2023-03-21 13:40 /a3
drwxr-xr-x   - hadoop supergroup          0 2023-03-17 11:35 /app
-rw-r--r--   3 root   supergroup     194000 2023-03-21 14:01 /hdfsbasic
-rw-r--r--   3 root   supergroup      60035 2023-03-21 14:01 /hdfsdata
drwx-wx-wx   - hadoop supergroup          0 2023-03-16 12:52 /hive
drwx-wx-wx   - hadoop supergroup          0 2023-03-16 12:53 /tmp
[hadoop@server201 ~]$
```

图 11-5　Hadoop 平台文件目录

11.3.3　Hadoop 数据清洗

将上传到 Hadoop 平台的酒店数据进行初步的数据清洗，使它符合大数据分析平台对数据的基本要求。

具体的清洗代码用 Java 程序编写，其中，酒店基本数据清洗及主程序的代码如下：

代码 11.3　HotelBasicClean.java

```
    public class HotelBasicClean extends Configured implements Tool {
        public static void main(String[] args) throws Exception {
            int run = ToolRunner.run(new HotelBasicClean(), new
String[]{"hdfs://192.168.56.201:8020/hdfsbasic","hdfs://192.168.56.201:8020/hot
elbasic"});
            System.exit(run);
        }
        public int run(String[] args) throws Exception {
            if (args.length < 2) {
                System.out.println("参数错误，使用方法：LineCharCountMR <Input>
<Output>");
                ToolRunner.printGenericCommandUsage(System.out);
                return 1;
            }
            Configuration config = getConf();
            config.set("fs.defaultFS", "hdfs://192.168.56.201:8020");
            // 这个解决 HDFS 问题
            config.set("fs.hdfs.impl",
org.apache.hadoop.hdfs.DistributedFileSystem.class.getName());
            // 这个解决本地 file 问题
            config.set("fs.file.impl",
org.apache.hadoop.fs.LocalFileSystem.class.getName());
            config.set("hadoop.home.dir", "/app/hdoop-3.2.3");
            // 通过如下的方式进行客户端身份的设置
            System.setProperty("HADOOP_USER_NAME", "root");
            System.setProperty("HADOOP_HOME", "/app/hdoop-3.2.3");

            FileSystem fs = FileSystem.get(config);
            Path dest = new Path(args[1]);
            if (fs.exists(dest)) {
                fs.delete(dest, true);
            }
            Job job = Job.getInstance(config, "LineChar");
            job.setJarByClass(getClass());
            job.setMapperClass(LineMapper.class);
```

```
            job.setOutputKeyClass(Text.class);
            job.setOutputValueClass(NullWritable.class);
            FileInputFormat.addInputPath(job, new Path(args[0]));
            FileOutputFormat.setOutputPath(job, dest);
            boolean b = job.waitForCompletion(true);
            return b ? 0 : 1;
        }
        //注意，最后一个参数为NullWritable，可以理解为NULL
        public static class LineMapper extends Mapper<LongWritable, Text, Text,
NullWritable> {
            Text ntxt=new Text();
            @Override
            protected void map(LongWritable key, Text value, Context context) throws
IOException, InterruptedException {
                String line = value.toString();
                if (StringUtils.isBlank(line)) {//删掉空行
                    return;
                }
                String[] arr=line.split(",");
                String a=arr[3];//地址
                //提取县市区替换地址
                if(a.contains("平度")) arr[3]= "平度市";
                else if(a.contains("市南")) arr[3]= "市南区";
                else if(a.contains("市北")) arr[3]= "市北区";
                else if(a.contains("李沧")) arr[3]= "李沧区";
                else if(a.contains("城阳")) arr[3]= "城阳区";
                else if(a.contains("崂山")) arr[3]= "崂山区";
                else if(a.contains("黄岛")) arr[3]= "黄岛区";
                else if(a.contains("即墨")) arr[3]= "即墨市";
                else if(a.contains("胶州")) arr[3]= "胶州市";
                else if(a.contains("莱西")) arr[3]= "莱西市";
                else arr[3] ="其他";
                //获取最后一列酒店详情
                String b=arr[5];
                if(b.equals("国家旅游局评定为四星级")) arr[5]="高档型";
                else if(b.equals("国家旅游局评定为五星级")) arr[5]="豪华型";
                else if(b.equals("国家旅游局评定为三星级")) arr[5]="舒适型";
                else if(b.equals("国家旅游局评定为二星级")) arr[5]="经济型";
                else;
                //新数组重新组成字符串，用逗号隔开
                String nline="";
                for (String ele:arr) {
                    nline+=ele+",";
                }
                nline=nline.substring(0,nline.length()-1);
                ntxt.set(nline);
                context.write(ntxt, NullWritable.get());
            }
        }
    }
```

酒店评论数据清洗的代码如下：

代码 11.4　HotelDataClean.java

```
    public class HotelDataClean extends Configured implements Tool {
        public static void main(String[] args) throws Exception {
            int run = ToolRunner.run(new HotelDataClean(), new
String[]{"hdfs://192.168.56.201:8020/hdfsdata",
"hdfs://192.168.56.201:8020/hoteldata"});
            System.exit(run);
        }
        public int run(String[] args) throws Exception {
            if (args.length < 2) {
                System.out.println("参数错误，使用方法：LineCharCountMR <Input>
<Output>");
                ToolRunner.printGenericCommandUsage(System.out);
                return 1;
            }
            Configuration config = getConf();
            config.set("fs.defaultFS", "hdfs://192.168.56.201:8020");
            config.set("fs.hdfs.impl", DistributedFileSystem.class.getName());
            config.set("fs.file.impl", LocalFileSystem.class.getName());
            System.setProperty("HADOOP_USER_NAME", "root");
            System.setProperty("HADOOP_HOME", "/app/hdoop-3.2.3");
            FileSystem fs = FileSystem.get(config);
            Path dest = new Path(args[1]);
            if (fs.exists(dest)) {
                fs.delete(dest, true);
            }
            Job job = Job.getInstance(config, "LineChar");
            job.setJarByClass(getClass());
            job.setMapperClass(LineMapper.class);
            job.setOutputKeyClass(Text.class);
            job.setOutputValueClass(NullWritable.class);
            FileInputFormat.addInputPath(job, new Path(args[0]));
            FileOutputFormat.setOutputPath(job, dest);
            boolean b = job.waitForCompletion(true);
            return b ? 0 : 1;
        }
        //注意，最后一个参数为 NullWritable，可以理解为 NULL
        public static class LineMapper extends Mapper<LongWritable, Text, Text,
NullWritable> {
            Text ntxt=new Text();
            @Override
            protected void map(LongWritable key, Text value, Context context) throws
IOException, InterruptedException {
                String line = value.toString();
                if (StringUtils.isBlank(line)) {//删掉空行
                    return;
                }
                String[] arr=line.split(",");

                //删掉最后一列
                String nline="";
                for (int i=0;i<arr.length-1;i++) {
                    nline+=arr[i]+",";
                }
```

```
            nline=nline.substring(0,nline.length()-1);
            ntxt.set(nline);
            context.write(ntxt, NullWritable.get());
        }
    }
}
```

清洗后的数据分别存放到 HDFS 的两个不同目录，hotelbasic 目录存放基本数据，hoteldata 目录存放评论数据，如图 11-6 所示。

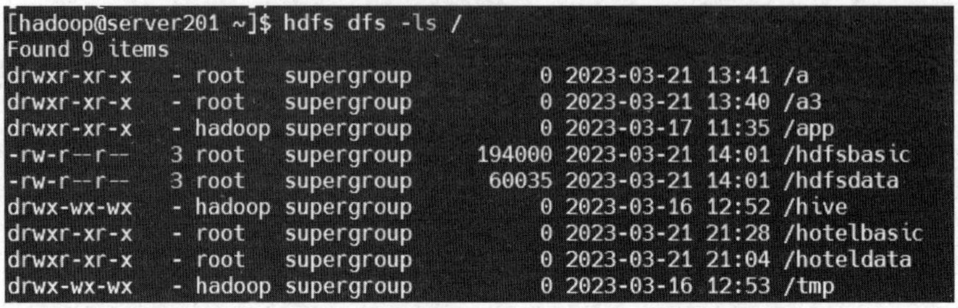

```
[hadoop@server201 ~]$ hdfs dfs -ls /
Found 9 items
drwxr-xr-x   - root   supergroup          0 2023-03-21 13:41 /a
drwxr-xr-x   - root   supergroup          0 2023-03-21 13:40 /a3
drwxr-xr-x   - hadoop supergroup          0 2023-03-17 11:35 /app
-rw-r--r--   3 root   supergroup     194000 2023-03-21 14:01 /hdfsbasic
-rw-r--r--   3 root   supergroup      60035 2023-03-21 14:01 /hdfsdata
drwx-wx-wx   - hadoop supergroup          0 2023-03-16 12:52 /hive
drwxr-xr-x   - root   supergroup          0 2023-03-21 21:28 /hotelbasic
drwxr-xr-x   - root   supergroup          0 2023-03-21 21:04 /hoteldata
drwx-wx-wx   - hadoop supergroup          0 2023-03-16 12:53 /tmp
```

图 11-6　存放清洗后数据的 Hadoop 文件系统目录

11.4　利用 MySQL 实现数据清洗

本节直接使用 MySQL 进行数据清洗，从而在数据分析全过程中仅使用 SQL 语句就能完成本项目的数据分析工作。

同样，我们直接使用第 2 章搭建好的伪分布式环境，其服务器节点的 IP 地址为 192.168.56.201，服务器名称为 server201。在 VisualBox 中启动 CentOS7-201 虚拟机，并使用 hadoop 账号登录系统。

10.4.1　hotelbasic.csv 数据清洗

hotelbasic.csv 文件的每一行数据是一个酒店的基本信息，按英文逗号分为酒店 id、酒店名称、酒店评分、酒店地址、酒店星级、星级详情。据此创建 6 个字段的表 hotelbasic，把 hotelbasic.csv 文件数据导入表中，命令如下：

```
    mysql> use weblog;
    mysql> create table hotelbasic(id char(8),name varchar(50),recommend double,
address  varchar(100),star varchar(20),stardetail varchar(50));
    mysql> load DATA infile '/tmp/hotelbasic.csv' into table hotelbasic
    character set utf8mb4 fields terminated by ',' lines terminated by '\n';
```

首先处理星级详情字段 stardetail，将所有的“国家旅游局评定为五星级”替换为“豪华型”，将“国家旅游局评定为四星级”替换为“高档型”，将“国家旅游局评定为三星级”替换为“舒适型”，将“国家旅游局评定为二星级”替换为“经济型”：

```
    mysql> update hotelbasic set stardetail='豪华型' where stardetail like '%五星
级%';
    mysql> update hotelbasic set stardetail='高档型' where stardetail like '%四星
级%';
```

```
    mysql> update hotelbasic set stardetail='舒适型' where stardetail like '%三星级%';
    mysql> update hotelbasic set stardetail='经济型' where stardetail like '%二星级%';
```

然后把酒店地址 address 中的详细地址改为区县名，具体如下：

```
    mysql> update hotelbasic set stardetail='市南区' where stardetail like '%市南%';
    mysql> update hotelbasic set stardetail='市北区' where stardetail like '%市北%';
    mysql> update hotelbasic set stardetail='李沧区' where stardetail like '%李沧%';
    mysql> update hotelbasic set stardetail='城阳区' where stardetail like '%城阳%';
    mysql> update hotelbasic set stardetail='崂山区' where stardetail like '%崂山%';
    mysql> update hotelbasic set stardetail='黄岛区' where stardetail like '%黄岛%';
    mysql> update hotelbasic set stardetail='即墨市' where stardetail like '%即墨%';
    mysql> update hotelbasic set stardetail='胶州市' where stardetail like '%胶州%';
    mysql> update hotelbasic set stardetail='莱西市' where stardetail like '%莱西%';
    mysql> update hotelbasic set stardetail='其他' where stardetail NOT REGEXP '
市南|市北|李沧|城阳|崂山|黄岛|即墨|胶州|莱西';
```

接下来把表 hotelbasic 中经过处理的数据导出为数据文件，具体如下：

```
    mysql>  select * from hotelbasic into outfile "hotelbasic.txt" fields terminated
by ',' lines terminated by '\n';
```

导出来的数据文件到/var/lib/mysql/weblog/目录下查找。数据文件先复制到 Linux 系统当前用户的主目录下，再执行下面操作保存到 HDFS 中，以方便导入 Hive 外部表进行处理：

```
    hdfs dfs -mkdir -p /hotelbasic
    hdfs dfs -put hotelbasic.txt /hotelbasic
```

10.4.2　hoteldata.csv 数据清洗

hoteldata.csv 文件的每一行数据是一个酒店用户对酒店的评分信息，按英文逗号分为酒店评论的用户名(user_name)、酒店名字(hotel_name)、出游类型(trip_type)、用户评价时间(evaluate_time)、酒店评论的发布日期(post_time)、酒店评论的用户打分(user_score)。据此创建 6 个字段的表 hoteldata，把 hoteldata.csv 文件数据导入表中，命令如下：

```
    mysql> use weblog;
    mysql> create table hoteldata(user_name char(20),hotel_name varchar(50),
trip_type char(12), evaluate_time varchar(20),post_time varchar(20),user_score
varchar(10));
    mysql> load DATA infile '/tmp/hoteldata.csv' into table hoteldata
    character set utf8mb4 fields terminated by ',' lines terminated by '\n';
```

导入数据后按清洗要求处理数据，此处略过。接下来把表 hoteldata 中经过数据清洗的数据导出为数据文件，具体如下：

```
    mysql>  select * from hoteldata into outfile "hoteldata.txt" fields terminated
by ',' lines terminated by '\n';
```

导出来的数据文件到/var/lib/mysql/weblog/目录下查找。数据文件先复制到 Linux 系统当前用户的主目录下，再执行下面操作保存到 HDFS 中，以方便导入 Hive 外部表进行处理：

```
hdfs dfs -mkdir -p /hoteldata
hdfs dfs -put hoteldata.txt /hoteldata
```

11.5 数据分析的实现

11.5.1 构建Hive数据仓库表

要使用酒店大数据分析服务平台，需要创建两张外部表，分别对应酒店基本信息表和酒店用户评论表。

5张Hive内部表中的数据分别是对酒店的5个关注角度进行数据分析得到的结果数据，内部表中的结果数据最终会导入MySQL中以方便用户查询。

各表结构设计如表11-1~表11-5所示。

表11-1 出游类型统计表设计

字段编号	字段名称	数据类型	约束	字段描述
1	triptype	varchar	PK	出游类型
2	num	int		数量

表11-2 用户满意度分布统计表设计

字段编号	字段名称	数据类型	约束	不是Null	字段描述
1	bad	Int		☑	不满意
2	good	Int		☑	满意
3	excellent	Int		☑	非常满意

表11-3 酒店星级分布统计表设计

字段编号	字段名称	数据类型	约束	不是Null	字段描述
1	stardetail	varchar	PK	☑	酒店星级类型
2	nums	Int		☑	数量

表11-4 酒店评论数据数量统计表设计

字段编号	字段名称	数据类型	约束	字段描述
1	hotel_name	varchar	PK	酒店星期
2	Num	int		数量

表11-5 各区县酒店数量和用户推荐评分统计表设计

字段编号	字段名称	数据类型	约束	字段描述
1	area_name	varchar	PK	区县名称
2	Num	int		数量
3	Recommend	Int		推荐平均评分

1. 基于上传的酒店用户评论数据 hoteldata.csv 创建 Hive 外部表

```
    create external table hotel_data(user_name string,hotel_name string,trip_type
string,time1 string,time2 string ,user_score double)
    ROW FORMAT DELIMITED FIELDS TERMINATED BY ',' LOCATION '/hoteldata';
```

这样就会自动加载/hoteldata 目录下面的数据。

2. 基于上传的酒店基本数据 hotelbasic.csv 创建 Hive 外部表

```
    create external table hotel_basic(id string,name string,recommend double,
address string,star string,stardetail string)
    ROW FORMAT DELIMITED FIELDS TERMINATED BY ',' LOCATION '/hotelbasic';
```

这样就会自动加载/hotelbasic 目录下面的数据。

3. 基于 Hive 外部表统计以下数据形成内部表，并将分析结果导出到 MySQL 数据库

（1）用户印象统计，用户对住过的酒店发表评论同时也进行打分，以下是根据用户对该地区或城市的酒店的总体评分情况来统计用户的总体印象。评分 4.5～5 分为优良，3.5～4.5 分为良好，3.5 分以下为差，用于统计酒店用户评分的等级比例。

提前建好 Hive 内部表 score_stat，并覆盖式插入数据到该表：

```
    create table score_stat(bad int,good int,excellent int);
```

插入数据：

```
    insert overwrite table score_stat select  count(case when user_score<3.5 then
1 else null end) as `bad`,count(case when user_score>=3.5 and user_score<4.5 then
1 else null end) as `good`,
    count(case when user_score>=4.5 then 1 else null end) as `excellent`  from
hotel_data;
```

注意：as 后的别名使用的是单撇号（`）。

（2）统计在线评论数最多的十大酒店，即十大网络人气酒店。

一般情况下，一家酒店的评论数量能代表这家酒店的人气，这里统计的是酒店名称和评论数目前十的酒店，设计评论数目表结构：

```
    create table comments_stat(hotel_name string,nums int);
    insert overwrite table comments_stat select hotel_name,count(1) as nums from
hotel_data group by hotel_name;
```

（3）不同旅游类型占比统计：

```
    create table triptype_stat(triptype string,nums bigint);
    insert overwrite table triptype_stat  select trip_type,count(1) as nums from
hotel_data group by trip_type;
```

（4）酒店星级分布情况统计。

设计酒店星级和数量结构：

```
    create table star_stat(stardetail string,nums int);
    insert overwrite table star_stat  select stardetail,count(1) from hotel_basic
```

```
group by stardetail;
```

（5）城市不同区的酒店数量分布情况，以热力图方式呈现：

```
create table area_stat(area_name string,nums int,recommend int);
```

插入数据：

```
insert overwrite table area_stat  select address, count(1),
round(avg(recommend), 2) from hotel_basic group by address;
```

以上查询语句是使用正则表达式从酒店数据的酒店地址中提取区县的名称。

读者可以分别执行以上语句，根据操作步骤得到分析结果；也可以将以上命令制成shell脚本，然后命名为bigdata.sh，通过运行脚本来构建数据仓库并取得分析结果。bigdata.sh完整代码如下：

代码 11.5 bigdata.sh

```
#!/bin/sh
#创建Hive外部表
echo "基于上传的hoteldata.csv创建Hive外部表"
hive -e "create external table if not exists  hotel_data(user_name
string,hotel_name string,trip_type string,time1 string,time2 string ,user_score
double)
ROW FORMAT DELIMITED FIELDS TERMINATED BY  ',' LOCATION '/hoteldata'"
echo "基于上传的hotelbasic.csv创建Hive外部表"
hive -e "create external table if not exists  hotel_basic(id string,name
string,recommend double,address string,star  string,stardetail string)
ROW FORMAT DELIMITED FIELDS TERMINATED BY ',' LOCATION '/hotelbasic'"
#创建Hive内部表
echo "==========================创建Hive内部表=========================="
#1.来该地区的用户的旅游类型统计
echo "1.用户的旅游类型统计表......................................"
hive -e "create table triptype_stat(triptype string,nums bigint)"
hive -e "insert overwrite table triptype_stat  select trip_type,count(1) as nums
from hotel_data group by trip_type"
#2.根据用户对该地区或城市的酒店的总体评分情况来统计用户的总体印象，用户评分统计
echo "2.根据用户对该地区或城市的酒店的总体评分情况来统计用户的总体印
象......................................"
hive -e "create table score_stat(bad int,good int,excellent int)"
hive -e "insert overwrite table score_stat select  count(case when
user_score<3.5 then 1 else null end)  as `bad`,count(case when user_score>=3.5 and
user_score<4.5 then 1 else null end)  as `good`,count(case when user_score>=4.5 then
1 else null end)  as `excellent`  from hotel_data"
#3.十大网络人气酒店，酒店名称和用户评论数量
echo "3.十大网络人气酒店......................................"
hive -e "create table comments_stat(hotel_name string,nums int)"
hive -e "insert overwrite table comments_stat select hotel_name,count(1) as nums
from hotel_data group by hotel_name"
#4.酒店星级分布情况统计，设计酒店星级和数量结构
echo "4.酒店星级分布情况统计......................................"
hive -e "create table star_stat(stardetail string,nums int)"
hive -e "insert overwrite table star_stat  select stardetail,count(1)  from
hotel_basic group by stardetail"
#5.各地区酒店数量统计
```

```
    echo "各地区酒店数量统计.........................................."
    hive -e "create table area_stat(area_name string,nums int,recommend int)"
    hive -e "insert overwrite table area_stat  select
address ,count(1),round(avg(recommend),2)  from hotel_basic group by  address"

    echo "分析结果已经保存到 Hive 内部表................................."
```

运行过程（部分截图）如图 11-7 所示。

```
[hadoop@server201 ~]$ ./bigdata.sh
基于上传的hoteldata.csv创建hive外部表
Hive Session ID = 74216407-eebb-490b-a5d6-3fb8cb7666a7

Logging initialized using configuration in jar:file:/app/hive-3.1.3/lib/hive-common-3.1.3.jar!/hive-log4j2.properties
Hive Session ID = 5399613a-514a-453e-b24f-bb97f9a84436
OK
Time taken: 2.199 seconds
WARN: The method class org.apache.commons.logging.impl.SLF4JLogFactory#release() was invoked.
WARN: Please see http://www.slf4j.org/codes.html#release for an explanation.
基于上传的hotelbasic.csv创建hive表
Hive Session ID = 49a5d0f1-6114-4407-a3c4-487eb649b6bc

Logging initialized using configuration in jar:file:/app/hive-3.1.3/lib/hive-common-3.1.3.jar!/hive-log4j2.properties
Hive Session ID = bbcc4720-d7a2-4f89-bb93-3d1ab7147810
OK
Time taken: 1.769 seconds
WARN: The method class org.apache.commons.logging.impl.SLF4JLogFactory#release() was invoked.
WARN: Please see http://www.slf4j.org/codes.html#release for an explanation.
===============================创建hive内部表===============================
1.用户的旅游类型统计表.........................................
Hive Session ID = 29b720ba-699f-4a4c-a750-e09cf05d2b10

Logging initialized using configuration in jar:file:/app/hive-3.1.3/lib/hive-common-3.1.3.jar!/hive-log4j2.properties
Hive Session ID = ddc90a38-60a4-4f1c-9971-ef68ecd9ee00
OK
Time taken: 2.023 seconds
WARN: The method class org.apache.commons.logging.impl.SLF4JLogFactory#release() was invoked.
WARN: Please see http://www.slf4j.org/codes.html#release for an explanation.
Hive Session ID = bfbd9f62-3d1c-4266-82c7-194603c2f60b

Logging initialized using configuration in jar:file:/app/hive-3.1.3/lib/hive-common-3.1.3.jar!/hive-log4j2.properties
Hive Session ID = 4ab5bdf9-99f5-4463-b4f1-6a038df95cef
Query ID = hadoop_20230323100042_c3649104-2936-4e17-8dd2-13d41ca30b4e
Total jobs = 2
Launching Job 1 out of 2
Number of reduce tasks not specified. Estimated from input data size: 1
In order to change the average load for a reducer (in bytes):
```

图 11-7　构建 Hive 数据表脚本的执行过程

通过运行以上脚本，根据 Hive 数据仓库外部表数据，对来该地区的用户的旅游类型、城市酒店数据评分（用户印象）、十大网络人气酒店、各区县酒店星级分布情况、各地区酒店数量及平均推荐指数等指标进行数据分析，每一项统计结果都构建 Hive 内部表。

11.5.2　导出结果数据到 MySQL

接下来，需要把 Hive 统计结果数据导入 MySQL 中，这样大数据应用开发人员才可以开发相应的查询系统供用户使用。MySQL 数据库对应的库和表结构需要提前创建好，以对应 Hive 内部表的统计结果数据。MySQL 中建库和表的脚本如下：

```
    create database hotel;
    create table triptype_stat(triptype varchar(33),nums int);
    create table score_stat(bad int,good int,excellent int);
    create table comments_stat(hotel_name varchar(200),nums int);
    create table star_stat(stardetail varchar(200),nums int);
    create table area_stat(area_name varchar(200),nums int,recommend
decimal(5,2));
```

1. 利用文件导入数据到 MySQL

MySQL 中保存统计结果数据的表建好后，可以把 Hive 表中的统计结果数据导出为数据文件，再把数据文件导入 MySQL 数据库对应的表中。从 Hive 的 triptype_stat 表中导出数据的语句如下：

```
hive> insert overwrite local directory '/tmp/triptype_stat' ROW FORMAT DELIMITED
FIELDS TERMINATED by ',' select * from triptype_stat;
```

语句执行结果如图 11-8 所示。

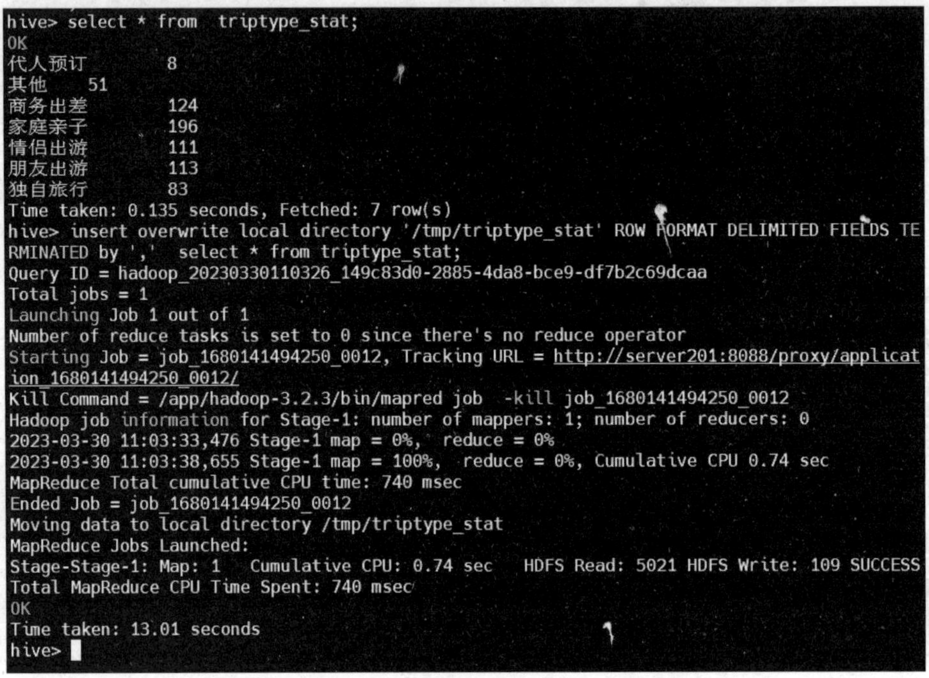

```
hive> select * from  triptype_stat;
OK
代人预订        8
其他    51
商务出差        124
家庭亲子        196
情侣出游        111
朋友出游        113
独自旅行        83
Time taken: 0.135 seconds, Fetched: 7 row(s)
hive> insert overwrite local directory '/tmp/triptype_stat' ROW FORMAT DELIMITED FIELDS TE
RMINATED by ','  select * from triptype_stat;
Query ID = hadoop_20230330110326_149c83d0-2885-4da8-bce9-df7b2c69dcaa
Total jobs = 1
Launching Job 1 out of 1
Number of reduce tasks is set to 0 since there's no reduce operator
Starting Job = job_1680141494250_0012, Tracking URL = http://server201:8088/proxy/applicat
ion_1680141494250_0012/
Kill Command = /app/hadoop-3.2.3/bin/mapred job  -kill job_1680141494250_0012
Hadoop job information for Stage-1: number of mappers: 1; number of reducers: 0
2023-03-30 11:03:33,476 Stage-1 map = 0%,  reduce = 0%
2023-03-30 11:03:38,655 Stage-1 map = 100%,  reduce = 0%, Cumulative CPU 0.74 sec
MapReduce Total cumulative CPU time: 740 msec
Ended Job = job_1680141494250_0012
Moving data to local directory /tmp/triptype_stat
MapReduce Jobs Launched:
Stage-Stage-1: Map: 1   Cumulative CPU: 0.74 sec   HDFS Read: 5021 HDFS Write: 109 SUCCESS
Total MapReduce CPU Time Spent: 740 msec
OK
Time taken: 13.01 seconds
hive>
```

图 11-8　从 Hive 的 triptype_stat 表中导出数据

导出的文件到相应的目录下查找。比如，执行上面命令把数据文件导出为/home/hadoop/triptype_stat/000000_0，读者可以自行查看文件内容来验证一下：

```
[hadoop@server201 hadoop]# cat /tmp/triptype_stat/000000_0
```

把数据文件导入 MySQL 对应的表中：

```
mysql> load DATA infile '/tmp/triptype_stat/000000_0' into table triptype_stat
character set utf8mb4 fields terminated by ',' lines terminated by '\n';
```

结果如图 11-9 所示。

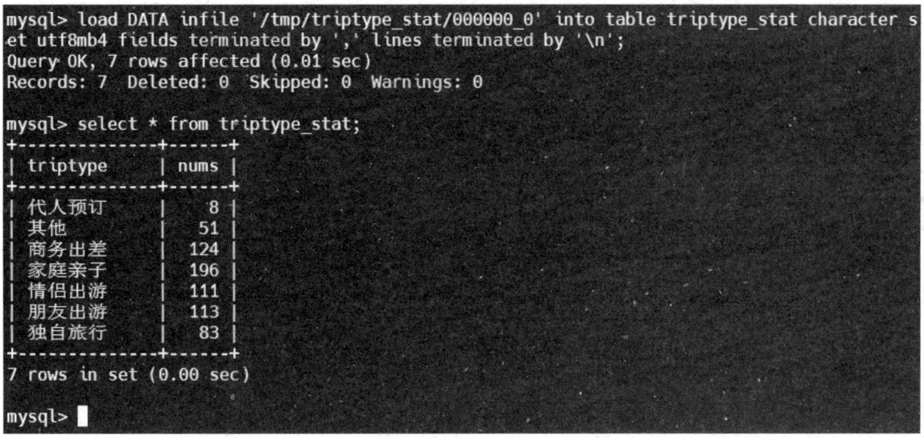

```
mysql> load DATA infile '/tmp/triptype_stat/000000_0' into table triptype_stat character s
et utf8mb4 fields terminated by ',' lines terminated by '\n';
Query OK, 7 rows affected (0.01 sec)
Records: 7  Deleted: 0  Skipped: 0  Warnings: 0

mysql> select * from triptype_stat;
+--------------+------+
| triptype     | nums |
+--------------+------+
| 代人预订     |    8 |
| 其他         |   51 |
| 商务出差     |  124 |
| 家庭亲子     |  196 |
| 情侣出游     |  111 |
| 朋友出游     |  113 |
| 独自旅行     |   83 |
+--------------+------+
7 rows in set (0.00 sec)

mysql> 
```

图 11-9　把数据文件导入 MySQL 对应的表中

Hive 中其他统计结果表的数据也可以这样导入 MySQL 数据库对应的表中，请读者自行完成。

2. 利用 Java 编程导入数据到 MySQL

创建好 MySQL 表之后，采用 Java 代码读取 Hive 表中的数据，然后将读取的结果集提取出来并通过 JDBC 的方式插入 MySQL 中。这里以 triptype_stat 表数据的导入为例来进行讲解，其他表的操作类似。

步骤 01 首先在 server201 主机上启动 Hadoop 集群，并启动 hiveserver2（参见 2.9 节）：

```
hadoop@server201 hadoop$ bin/hiveserver2
```

步骤 02 编写 Java 代码实现数据读取，并通过 JDBC 方式将数据插入 MySQL，该步骤需要将 MySQL 的 JDBC 驱动包导入项目。实现的具体代码如下：

```
import java.sql.*;
import java.sql.SQLException;
public class HivetoMySQL {
    private static String driverName = "org.apache.hive.jdbc.HiveDriver";
    private static String driverName_mysql = "com.mysql.jdbc.Driver";
    public static void main(String[] args) throws SQLException {
        try {
            Class.forName(driverName);
        }catch (ClassNotFoundException e) {
            // TODO Auto-generated catch block
            e.printStackTrace();
            System.exit(1);
        }
        Connection con1 = DriverManager.getConnection("jdbc:hive2:
//192.168.56.201:10000/default", "hduser", "hduser");//后两个参数是用户名和密码
        if(con1 == null)
            System.out.println("连接失败");
        else {
            Statement stmt = con1.createStatement();
            String sql = "select * from triptype_stat";
            System.out.println("Running: " + sql);
```

```
                ResultSet res = stmt.executeQuery(sql);
                //InsertToMysql
                try {
                    Class.forName(driverName_mysql);
                    Connection con2 = DriverManager.getConnection("jdbc:mysql:
//192.168.56.201:3306/hotel","root","A1b2c3==");//MySQL 数据库用户名和密码
                    String sql2 = "insert into triptype_stat(triptype,nums) values
(?,?)";
                    PreparedStatement ps = con2.prepareStatement(sql2);
                    while (res.next()) {
                        ps.setString(1,res.getString(1));
                        ps.setString(2,res.getString(2));
                        ps.executeUpdate();
                    }
                    ps.close();
                    con2.close();
                    res.close();
                    stmt.close();
                } catch (ClassNotFoundException e) {
                    e.printStackTrace();
                }
            }
            con1.close();
        }
    }
```

通过以上操作实现了将 Hive 表 triptype_stat 的内容导入 MySQL 表 triptype_stat。其他的分析结果表的导入方式类似，请读者自行完成其他表的数据迁移。

11.6 分析结果数据可视化

到上一小节为止，数据分析的工作已经完成了。接下来的工作只是把上面的分析结果，即 MySQL 中的分析结果数据，以图表的方式呈现到 Web 网页上。这里采用 Spring Boot+MyBatis+MySQL 框架构建可视化项目，本节具体代码可以参见本书配套下载包中相关项目文件夹下的 HotelVisualization。

提示：如果读者没有 SMM 框架开发经验，可以尝试把分析结果数据导出为文本文件，再把数据复制到 Excel 中进行可视化处理，最后截图放在 Web 网页上展示。

11.6.1 数据可视化开发

图表展示使用 ECharts，ECharts 是一个纯 JavaScript 图表库，底层依赖于轻量级的 Canvas 类库 ZRender，基于 BSD 开源协议，是一款非常优秀的可视化前端框架。它提供了直观、生动、可交互、可高度个性化定制的数据可视化图表，其创新的拖曳重计算、数据视图、值域漫游等特性大大增强了用户体验，赋予用户对数据进行挖掘、整合的能力。ECharts 支持折线图（区域图）、柱状图（条状图）、散点图（气泡图）、K 线图、饼图（环形图）等。

步骤 01 在 IDEA 中新建一个 Maven 项目，引入需要使用的框架，如 Spring Boot、MyBatis 等。具体的 pom.xml 代码如下：

代码 11.6　pom.xml

```
    <?xml version="1.0" encoding="UTF-8"?>
    <project xmlns="http://maven.apache.org/POM/4.0.0"
xmlns:xsi="http://www.w3.org/2001/XMLSchema-instance"
xsi:schemaLocation="http://maven.apache.org/POM/4.0.0
http://maven.apache.org/xsd/maven-4.0.0.xsd">
        <modelVersion>4.0.0</modelVersion>
        <parent>
            <groupId>org.springframework.boot</groupId>
            <artifactId>spring-boot-starter-parent</artifactId>
            <version>2.1.6.RELEASE</version>
            <relativePath/> <!-- lookup parent from repository -->
        </parent>
        <groupId>com.zjp.echartsdemo</groupId>
        <artifactId>echartsdemo</artifactId>
        <version>0.0.1-SNAPSHOT</version>
        <name>echartsdemo</name>
        <description>Demo project for Spring Boot</description>

        <properties>
            <java.version>1.8</java.version>
        </properties>

        <dependencies>
            <dependency>
                <groupId>org.springframework.boot</groupId>
                <artifactId>spring-boot-starter-web</artifactId>
            </dependency>

            <dependency>
                <groupId>org.springframework.boot</groupId>
                <artifactId>spring-boot-starter-test</artifactId>
                <scope>test</scope>
            </dependency>
            <dependency>
                <groupId>org.webjars.bower</groupId>
                <artifactId>echarts</artifactId>
                <version>4.2.1</version>
            </dependency>
            <dependency>
                <groupId>org.webjars</groupId>
                <artifactId>jquery</artifactId>
                <version>3.4.1</version>
            </dependency>
            <dependency>
                <groupId>org.mybatis.spring.boot</groupId>
                <artifactId>mybatis-spring-boot-starter</artifactId>
                <version>2.0.1</version>
            </dependency>
            <dependency>
                <groupId>mysql</groupId>
```

```
            <artifactId>mysql-connector-java</artifactId>
            <scope>runtime</scope>
        </dependency>
        <dependency>
            <groupId>com.github.pagehelper</groupId>
            <artifactId>pagehelper-spring-boot-starter</artifactId>
            <version>1.2.5</version>
        </dependency>
        <!-- alibaba 的 druid 数据库连接池 -->
        <dependency>
            <groupId>com.alibaba</groupId>
            <artifactId>druid-spring-boot-starter</artifactId>
            <version>1.1.9</version>
        </dependency>
        <!-- 引入 Thymeleaf 依赖 -->
        <dependency>
            <groupId>org.springframework.boot</groupId>
            <artifactId>spring-boot-starter-thymeleaf</artifactId>
        </dependency>
        <dependency>
            <groupId>org.springframework.boot</groupId>
            <artifactId>spring-boot-devtools</artifactId>
            <optional>true</optional>
        </dependency>

    </dependencies>

    <build>
        <plugins>
            <plugin>
                <groupId>org.springframework.boot</groupId>
                <artifactId>spring-boot-maven-plugin</artifactId>
            </plugin>
        </plugins>
    </build>

</project>
```

步骤 02 配置 Spring Boot 的配置文件、数据库连接池以及 Web 服务器端口，具体代码如下：

代码 11.7 application.properties

```
server.port=8088
#数据库连接池配置
spring.datasource.name=zjptest
spring.datasource.type=com.alibaba.druid.pool.DruidDataSource
spring.datasource.druid.filters=stat
spring.datasource.druid.driver-class-name=com.mysql.jdbc.Driver
spring.datasource.druid.url=jdbc:mysql://192.168.56.201:3306/hotel?useUnicode=true&characterEncoding=UTF-8&allowMultiQueries=true&serverTimezone=UTC
spring.datasource.druid.username=root
spring.datasource.druid.password=123456
pring.datasource.druid.initial-size=1
spring.datasource.druid.min-idle=1
spring.datasource.druid.max-active=20
```

```
    spring.datasource.druid.max-wait=6000
    spring.datasource.druid.time-between-eviction-runs-millis=60000
    spring.datasource.druid.min-evictable-idle-time-millis=300000
    spring.datasource.druid.validation-query=SELECT 'x'
    spring.datasource.druid.test-while-idle=true
    spring.datasource.druid.test-on-borrow=false
    spring.datasource.druid.test-on-return=false
    spring.datasource.druid.pool-prepared-statements=false
    spring.datasource.druid.max-pool-prepared-statement-per-connection-size=20
    #MyBatis 配置
    mybatis.mapper-locations=classpath:mapper/*.xml
    mybatis.type-aliases-package= com.dpzhou.echartsdemo.echartsdemo.entity
    #Thymeleaf 配置
    spring.thymeleaf.cache=false
    spring.thymeleaf.prefix=classpath:/templates/
    spring.thymeleaf.suffix=.html
    spring.thymeleaf.mode=HTML5
    spring.thymeleaf.encoding=UTF-8
    spring.thymeleaf.check-template-location=true
```

步骤 03 根据项目需求，需要开发 5 个可视化统计图，因为项目使用了 MyBatis 框架，所以可以将 SQL 直接配置在 Mapper 文件中，详细代码如下：

代码 11.8 CommonMapper.xml

```
    <?xml version="1.0" encoding="UTF-8"?>
    <!DOCTYPE mapper PUBLIC "-//mybatis.org//DTD Mapper 3.0//EN"
"http://mybatis.org/dtd/mybatis-3-mapper.dtd">
    <mapper namespace="com.zjp.echartsdemo.echartsdemo.dao.CommonMapper">

        <select id="selectTripType" parameterType="java.lang.String"
resultType="map">
            select
             triptype,nums
            from triptype_stat order by nums desc limit 0,7
        </select>
        <select id="selectCommentsStat" parameterType="java.lang.String" resultType="map">
            select
             hotel_name,nums
            from comments_stat order by nums desc limit 0,10
        </select>
        <select id="selectScoreStat" parameterType="java.lang.String" resultType="map">
            select
             *
            from score_stat
         </select>
        <select id="selectStarStat" parameterType="java.lang.String"
resultType="map">
            select
             stardetail,nums
            from star_stat order by nums desc limit 0,4
        </select>
        <select id="selectAreaStat" parameterType="java.lang.String"
resultType="map">
           select * from area_stat where area_name in('市南区','市北区','黄岛区','即
```

```
墨区','城阳区','崂山区','李沧区') order by nums desc limit 0,10
        </select>
    </mapper>
```

前端异步请求数据采用 Ajax 技术，后台查询数据以 JSON 格式返回，通过 Echarts 在页面加载时返回 JSON 数据。

步骤 04 定义 view.html 页面，划分 div 分别显示对应的统计图表，页面 JavaScript 脚本采用 Ajax 实现无刷新交互，具体页面代码如下：

代码 11.9 view.html

```
    <!-- 为 ECharts 准备一个具备大小（宽高）的 DOM -->
    <h1 class="aa">青岛市酒店大数据可视化</h1>
    <div id="main" style="width:
500px;height:400px;position:absolute;top:100px"></div><!--用户的旅游类型统计-->
    <div id="main2" style="width:
500px;height:400px;position:absolute;top:100px;left:550px"></div><!--地区酒店数量
统计-->
    <div id="main3" style="width:
500px;height:400px;position:absolute;top:100px;left:1050px"></div><!--用户印象评
分等级比例统计-->
    <a href="/echarts" >Goto 基于用户评论统计分析可视化</a>
    <script type="text/javascript">
        // 基于准备好的 DOM，初始化 ECharts 实例
        var myChart = echarts.init(document.getElementById('main'));
        // 新建 nums 数组来接收数据
        var triptypes = [];
        var nums = [];
        var json = {};
        var datatemp = [];

        //旅游类型统计
        $.ajax({
            type:"GET",
            url:"/triptypestat",
            dataType:"json",
            async:false,
            success:function (result) {
                json = result;
                for (var i = 0; i < result.length; i++){
                    triptypes.push(result[i].triptype);
                    nums.push(result[i].nums);
                    var ob = {name:"",value:""};
                    ob.name = result[i].triptype;
                    ob.value = result[i].nums;
                    datatemp.push(ob);
                }


            },
            error :function(errorMsg) {
                alert("获取后台数据失败！");
            }
```

```
});

// 指定图表的配置项和数据
var option = {
    title: {
        text: '用户的旅游类型统计'
    },
    tooltip: {},
    legend: {
        data:['人次']
    },
    xAxis: {
        //结合
        axisLabel: {
            interval:0,
            rotate:20
        },
        data: triptypes
    },

    yAxis: {},
    series: [{
        name: '旅游类型',
        type: 'bar',
        //结合
        data: nums
    }]
};

// 使用刚指定的配置项和数据显示图表
myChart.setOption(option);

//加载地图数据
//旅游类型统计
$.ajax({
    type:"GET",
    url:"/areastat",
    dataType:"json",
    async:false,
    success:function (result) {
        json = result;
        for (var i = 0; i < result.length; i++){
            var ob = {name:"",value:""};
            ob.name = result[i].area_name;
            ob.value = result[i].nums;
            datatemp.push(ob);
        }

    },
    error :function(errorMsg) {
        alert("获取后台数据失败! ");
    }
});
var myChart2 = echarts.init(document.getElementById('main2'));
```

```
option = {
    title: {
        text: '各地区酒店数量统计'
    },
    tooltip: {
        formatter:function(params,ticket, callback){
            return params.seriesName+'<br />'+params.name+': '+params.value
        }
    },

    visualMap: {
        min: 0,
        max: 1500,
        left: 'left',
        top: 'bottom',
        text: ['高','低'],
        inRange: {
            color: ['#e5e0e0', '#490104']
        },
        show:true
    },
    geo: {
        map: 'QD',
        roam: false,
        zoom:1.23,
        label: {
            normal: {
                show: true,
                fontSize:'10',
                color: 'rgba(0,0,0,0.7)'
            }
        },
        itemStyle: {
            normal:{
                borderColor: 'rgba(0, 0, 0, 0.2)'
            },
            emphasis:{
                areaColor: '#F3B329',
                shadowOffsetX: 0,
                shadowOffsetY: 0,
                shadowBlur: 20,
                borderWidth: 0,
                shadowColor: 'rgba(0, 0, 0, 0.5)'
            }
        }
    },

    series : [
        {
            name: '酒店数量',
            type: 'map',
            geoIndex: 0,
            data:datatemp
        }
```

```
    ]
};

$.getJSON('js/370200.json', function (geoJson) {
    myChart2.hideLoading();
    echarts.registerMap('QD', geoJson);
    myChart2.setOption(option);
})

//酒店星级分布情况统计
var myChart3 = echarts.init(document.getElementById('main3'));
var option3 = {
    title : {
        text: '酒店星级分布情况统计',
        subtext: '',
        x:'center'
    },
    tooltip : {
        trigger: 'item',
        formatter: "{a} <br/>{b} : {c} ({d}%)"
    },
    legend: {
        orient: 'vertical',
        left: 'left',
    },
    series : [
        {
            name: '酒店类型',
            type: 'pie',
            radius : '55%',
            center: ['50%', '60%'],
            data:(function () {

                var datas = [];
                $.ajax({
                    type:"POST",
                    url:"/starstat",
                    dataType:"json",
                    async:false,
                    success:function (result) {
                        for (var i = 0; i < result.length; i++){
                            datas.push({
                                "value":result[i].nums,
"name":result[i].stardetail
                            })
                        }
                    }
                })
                return datas;
            })(),
            itemStyle: {
                emphasis: {
                    shadowBlur: 10,
                    shadowOffsetX: 0,
                    shadowColor: 'rgba(0, 0, 0, 0.5)'
```

```
                    }
                }
            }
        ]
    };
    myChart3.setOption(option3);
</script>
```

11.6.2 数据可视化部署

Spring Boot 中 Controller、Service 和 Dao 层的代码不在本书讨论范围之内，具体可以参见本书配套的项目源码。

接下来讨论一下数据可视化开发完成后，项目 HotelVisualization 部署的详细过程。可以按照如下步骤改造项目中的部分文件，将项目部署到外接的 Tomcat 服务器中。

步骤 01 改造 Spring Boot 项目。

以 HotelVisualization 项目为例，首先修改 pom.xml 中的 packaging 目标为 war，并将内置 Tomcat 排除，引入外部 Tomcat 依赖 spring-boot-starter-tomcat：

```
    <?xml version="1.0" encoding="UTF-8"?>
    <project xmlns="http://maven.apache.org/POM/4.0.0"
xmlns:xsi="http://www.w3.org/2001/XMLSchema-instance"
         xsi:schemaLocation="http://maven.apache.org/POM/4.0.0
http://maven.apache.org/xsd/maven-4.0.0.xsd">
        <modelVersion>4.0.0</modelVersion>
        <parent>
            <groupId>org.springframework.boot</groupId>
            <artifactId>spring-boot-starter-parent</artifactId>
            <version>2.1.6.RELEASE</version>
            <relativePath/> <!-- lookup parent from repository -->
        </parent>
        <groupId>com.zjp.echartsdemo</groupId>
        <artifactId>echartsdemo</artifactId>
        <version>0.0.1-SNAPSHOT</version>
        <name>echartsdemo</name>
       <packaging>war</packaging>
        <description>Demo project for Spring Boot</description>

        <properties>
            <java.version>1.8</java.version>
        </properties>

        <dependencies>
            <dependency>
                <groupId>org.springframework.boot</groupId>
                <artifactId>spring-boot-starter-web</artifactId>
               <exclusions>
                <exclusion>
                    <groupId>org.springframework.boot</groupId>
                    <artifactId>spring-boot-starter-tomcat</artifactId>
                </exclusion>
                </exclusions>
```

```
    </dependency>
    <!-- 添加外部 Tomcat 包 -->
    <dependency>
      <groupId>org.springframework.boot</groupId>
      <artifactId>spring-boot-starter-tomcat</artifactId>
      <!-- 但是这里一定要设置为 provided -->
      <scope>provided</scope>
    </dependency>
    <dependency>
       <groupId>org.springframework.boot</groupId>
       <artifactId>spring-boot-starter-test</artifactId>
       <scope>test</scope>
    </dependency>
    <dependency>
       <groupId>org.webjars.bower</groupId>
       <artifactId>echarts</artifactId>
       <version>4.2.1</version>
    </dependency>
    <dependency>
       <groupId>org.webjars</groupId>
       <artifactId>jquery</artifactId>
       <version>3.4.1</version>
    </dependency>
    <dependency>
       <groupId>org.mybatis.spring.boot</groupId>
       <artifactId>mybatis-spring-boot-starter</artifactId>
       <version>2.0.1</version>
    </dependency>
    <dependency>
       <groupId>mysql</groupId>
       <artifactId>mysql-connector-java</artifactId>
       <scope>runtime</scope>
    </dependency>
    <dependency>
       <groupId>com.github.pagehelper</groupId>
       <artifactId>pagehelper-spring-boot-starter</artifactId>
       <version>1.2.5</version>
    </dependency>
    <!-- alibaba 的 druid 数据库连接池 -->
    <dependency>
       <groupId>com.alibaba</groupId>
       <artifactId>druid-spring-boot-starter</artifactId>
       <version>1.1.9</version>
    </dependency>
    <!-- 引入 Thymeleaf 依赖 -->
    <dependency>
       <groupId>org.springframework.boot</groupId>
       <artifactId>spring-boot-starter-thymeleaf</artifactId>
    </dependency>
    <dependency>
       <groupId>org.springframework.boot</groupId>
       <artifactId>spring-boot-devtools</artifactId>
       <optional>true</optional>
    </dependency>
</dependencies>
```

```
    <build>
        <plugins>
            <plugin>
                <groupId>org.springframework.boot</groupId>
                <artifactId>spring-boot-maven-plugin</artifactId>
            </plugin>
        </plugins>
    </build>

</project>
```

然后修改程序入口类 EchartsdemoApplication.java，使它继承自 SpringBootServletInitializer，并实现 configurer 方法：

```
package com.zjp.echartsdemo.echartsdemo;

import org.mybatis.spring.annotation.MapperScan;
import org.springframework.boot.SpringApplication;
import org.springframework.boot.autoconfigure.SpringBootApplication;

@SpringBootApplication
@MapperScan(basePackages = {"com.dpzhou.echartsdemo.echartsdemo.dao"})
public class EchartsdemoApplication extends SpringBootServletInitializer {

    public static void main(String[] args) {
        SpringApplication.run(EchartsdemoApplication.class, args);
    }

    /**
     * 重写 configure 方法
     */
    @Override
    protected SpringApplicationBuilder configure(SpringApplicationBuilder 
builder) {
        return builder.sources(EchartsdemoApplication.class);
    }
}
```

最后使用“mvn clean package”执行构建，将在 target 目录下生成“echartsdemo-0.0.1-SNAPSHOT.war”文件。

接下来，介绍一下如何把这个可视化项目部署到本地 Linux 服务器（即主机名为 server201 的这台机器）的 Tomcat 容器上。

步骤 02 下载安装 Tomcat。

使用 wget 命令连接到 Apache 下载服务器，下载 Tomcat 9.0.73 压缩包：

```
root@server201:~# wget 
https://downloads.apache.org/tomcat/tomcat-9/v9.0.73/bin/apache-tomcat-9.0.73.t
ar.gz
```

将 Tomcat 压缩包解压，并把解压出来的目录修改为/opt/tomcat 目录：

```
root@server201:~# tar -zxvf apache-tomcat-9.0.73.tar.gz
```

```
root@server201:~# mv apache-tomcat-9.0.73 /opt/tomcat
```

步骤 03　上传 war 包。

使用 MobaXterm 将 echartsdemo-0.0.1-SNAPSHOT.war 上传到服务器，然后将它移动到/opt/tomcat/ webapps/目录下：

```
root@server201:~# mv echartsdemo-0.0.1-SNAPSHOT.war /opt/tomcat/webapps/
```

启动 Tomcat：

```
root@server201:~# sh /opt/tomcat/bin/startup.sh
Using CATALINA_BASE:   /opt/tomcat
Using CATALINA_HOME:   /opt/tomcat
Using CATALINA_TMPDIR: /opt/tomcat/temp
Using JRE_HOME:        /usr/local/java/jdk1.8.0_121/jre
Using CLASSPATH:
/opt/tomcat/bin/bootstrap.jar:/opt/tomcat/bin/tomcat-juli.jar
Using CATALINA_OPTS:
Tomcat started.
```

通过浏览器访问 http://192.168.56.201:8080/，查看 Tomcat 运行状态，如图 11-10 所示。

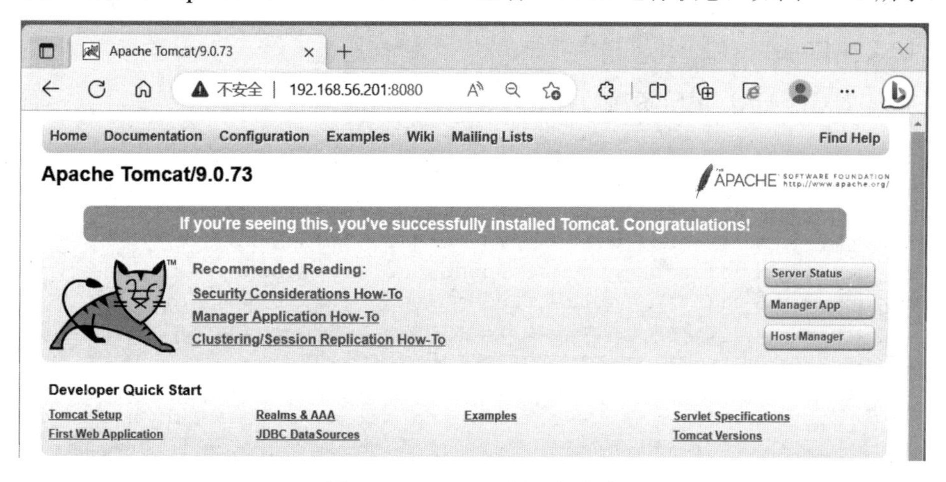

图 11-10　Tomcat 的运行状态

出现 Tomcat 介绍页说明 Tomcat 启动成功了，而我们的 echartsdemo-0.0.1-SNAPSHOT.war 包会被 Tomcat 自动解包放置在 webapps 目录下。

步骤 04　配置 Tomcat。

修改/opt/tomcat/conf/server.xml 文件，在<Host>节点中添加<Context>配置：

```
root@server201:~# sudo vim /opt/tomcat/conf/server.xml
…
<Server>
…
<Service>
…
<Engine name="Catalina" defaultHost="192.168.56.201">
…
```

```
        <Host name="192.168.56.201"  appBase="webapps"
            unpackWARs="true" autoDeploy="true">
          <!--
           …
          -->
          <Valve className="org.apache.catalina.valves.AccessLogValve"
    directory="logs"
                prefix="localhost_access_log" suffix=".txt"
                pattern="%h %l %u %t &quot;%r&quot; %s %b" />
          <Context path="/echartsdemo" docBase="echartsdemo-0.0.1-SNAPSHOT"
reloadable="reloadable"/>
        </Host>
    </Engine>
    </Services>
    </Server>
```

其中<Context>配置的 path 填写需要映射的上下文路径，docBase 填写项目的名字（war 包的名字去除.war 后缀）。修改完成后，保存并重新启动 Tomcat。

对于上面的配置，echartsdemo 项目将会运行在 http://192.168.56.201:8080/echartsdemo 这个路径上。

至此，构建 war 包并部署在 Tomcat 容器上的工作就已完成。接下来，可以根据不同的统计分析对应的 URL 来访问不同的统计可视化界面。

步骤 05 根据不同的 URL 访问不同的统计可视化界面。

具体 URL 与统计分析功能的对应关系参照 CommonController，代码如下：

```
package com.zjp.echartsdemo.echartsdemo.controller;
import com.zjp.echartsdemo.echartsdemo.dao.CommonMapper;
import com.zjp.echartsdemo.echartsdemo.dao.TripTypeMapper;
import com.zjp.echartsdemo.echartsdemo.entity.TripType;
import org.springframework.beans.factory.annotation.Autowired;
import org.springframework.stereotype.Controller;
import org.springframework.web.bind.annotation.RequestMapping;
import org.springframework.web.bind.annotation.ResponseBody;
import java.util.ArrayList;
import java.util.HashMap;
import java.util.List;
import java.util.Map;
@Controller
public class CommonController {
    @Autowired
    CommonMapper commonMapper;
    @RequestMapping("/triptypestat")
    @ResponseBody
    public List<Map> triptypestat() {
        List<Map> lst = new ArrayList<Map>();
        lst = commonMapper.selectTripType();
        return lst;
    }
    /**
     * 查询区域酒店数量
     * @return
```

```
     */
    @RequestMapping("/areastat")
    @ResponseBody
    public List<Map> areastat() {
        List<Map> lst = new ArrayList<Map>();
        lst = commonMapper.selectAreaStat();
        return lst;
    }
    /**
     * 用户满意度评论比例
     * @return
     */
    @RequestMapping("/commentstat")
    @ResponseBody
    public List<Map> commentstat() {
        List<Map> lst = new ArrayList<Map>();
        lst = commonMapper.selectCommentsStat();
        return lst;
    }
    /**
     * 用户十大人气  十大网络人气酒店
     * @return
     */
    @RequestMapping("/scorestat")
    @ResponseBody
    public List<Map> scorestat() {
        List<Map> lst = new ArrayList<Map>();
        lst = commonMapper.selectScoreStat();
        List<Map> rlst = new ArrayList<Map>();//存储用户评分等级及酒店数量
        if(lst!=null&&lst.size()>0){
            Map map = lst.get(0);
            Map mm=new HashMap();
            mm.put("level","非常满意");
            mm.put("nums",Integer.parseInt(map.get("excellent").toString()));
            rlst.add(mm);
            Map mm1=new HashMap();
            mm1.put("level","满意");
            mm1.put("nums",Integer.parseInt(map.get("good").toString()));
            rlst.add(mm1);
            Map mm2=new HashMap();
            mm2.put("level","不满意");
            mm2.put("nums",Integer.parseInt(map.get("bad").toString()));
            rlst.add(mm2);
        }
        return rlst;
    }
    /**
     * 酒店星级统计
     * @return
     */
    @RequestMapping("/starstat")
    @ResponseBody
    public List<Map> starstat() {
        List<Map> lst = new ArrayList<Map>();
        lst = commonMapper.selectStarStat();
```

```
        return lst;
    }
}
```

其中，基于酒店基本数据的统计分析部分包括用户旅游类型统计、各地区酒店数量统计、酒店星级情况统计，可视化结果如图 11-11 所示。

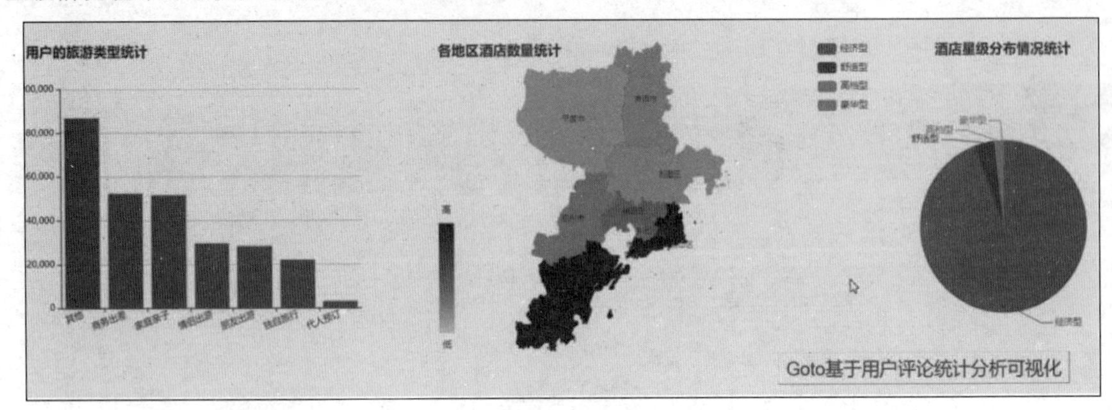

图 11-11　酒店基本数据统计图

基于酒店用户评论数据的统计分析部分包括十大网络人气酒店和用户满意度统计，可视化结果如图 11-12 所示。

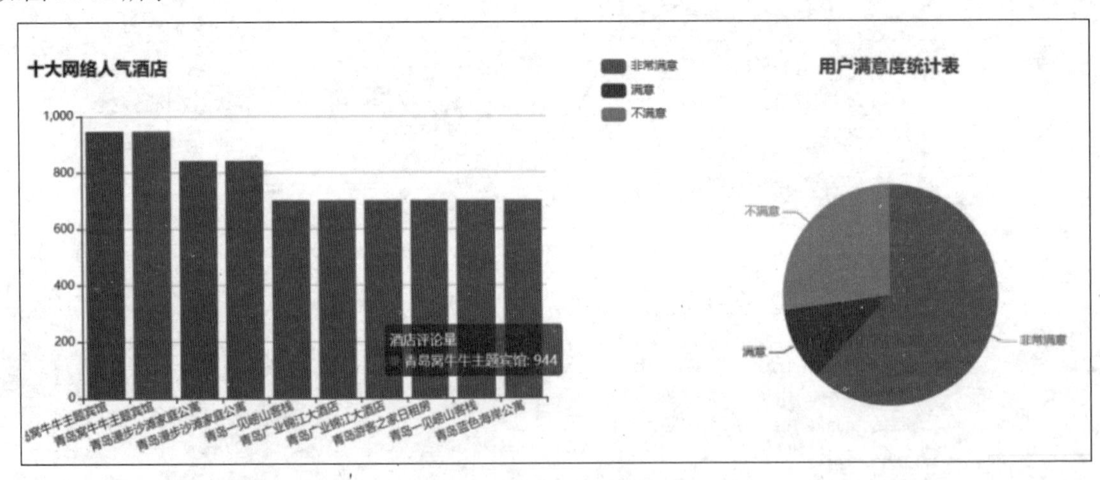

图 11-12　用户评论数据统计图